STP
cs 7A

MAR 2000

)9

STP
NATIONAL CURRICULUM
MATHEMATICS
7A

L. BOSTOCK, B.Sc.

S. CHANDLER, B.Sc.

A. SHEPHERD, B.Sc.

E. SMITH, M.Sc.

STANLEY THORNES (PUBLISHERS) LTD

First published in 1995 by
Stanley Thornes (Publishers) Ltd,
Ellenborough House,
Wellington Street,
CHELTENHAM GL50 1YD

A catalogue record of this book is available from the British Library.

ISBN 0–7487–2005–7

Artwork by Linda Jeffrey, Jeff Edwards

Front cover image produced using material kindly supplied by I LOVE LOVE CO, makers of the The Happy Cube © Laureyssens/Creative City Ltd 1986/91. Distributed in UK by: RIGHTRAC, 119 Sandycombe Road, Richmond, Surrey TW9 2ER Tel. 0181 940 3322

The publishers are grateful to the following for granting permission to reproduce copyright material:
Central Statistical Office: pp. 46, 48, 419, 424.
Argos: p. 371.

Typeset by Tech-Set, Gateshead, Tyne & Wear.
Printed and bound in Great Britain at T J Press, Padstow, Cornwall

CONTENTS

INTRODUCTION

To the pupil

This is the first book of a series that helps you to learn, enjoy and progress through Mathematics in the National Curriculum. As well as a clear and concise text the book offers a wide range of practical and investigational work that is relevant to the mathematics you are learning.

Everyone needs success and satisfaction in getting things right. With this in mind we have divided many of the exercises into three types.

The first type, identified by plain numbers, e.g. **15**, helps you to see if you understand the work. These questions are considered necessary for every chapter.

The second type, identified by an underline, e.g. **15**, are extra, but not harder, questions for quicker workers, for extra practice or for later revision.

The third type, identified by a coloured square, e.g. **15** , are for those of you who like a greater challenge.

Most chapters have a 'mixed exercise' after the main work of the chapter has been completed. This will help you to revise what you have done, either when you have finished the chapter or at a later date. All chapters end with some mathematical puzzles or practical or investigational work. For this work you are encouraged to share your ideas with others, to use any mathematics you are familiar with, and to try to solve each problem in different ways, appreciating the advantages and disadvantages of each method.

After every four or five chapters you will find a Summary. This lists the most important points that have been studied in the previous chapters and concludes with revision exercises that test the work you have studied up to that point.

All of you need to be able to use a calculator accurately but it is unwise to rely on a calculator for work that you should do in your head. By all means use it to check your calculations, but remember, whether you use a calculator or do the working yourself, always estimate your answer and always ask yourself the question, 'Is my answer a sensible one?'

Mathematics is an exciting and enjoyable subject when you understand what is going on. Remember, if you don't understand something, ask someone who can explain it to you. If you still don't understand, ask again. Good luck with your studies.

To the teacher

This is the first book of the series STP National Curriculum Mathematics. It is based on the ST(P) Mathematics series but has been extensively rewritten and is now firmly based on the Programme of Study for Key Stages 3 and 4, starting at Level 4.

The A series of books aims to prepare pupils for about Level 8 at Key Stage 3 and for the higher tier at GCSE.

ADDITION AND SUBTRACTION OF WHOLE NUMBERS

We use whole numbers all the
time in everyday life.
Consider this example.
Jane takes a bar of chocolate
costing 37 p and a card priced at
86 p to the cash desk.
The cashier asks for £1.43.
Jane might react by

- paying the amount asked
  because she has no idea what
  the sum of 37 p and 86 p should be
- asking the cashier to check the amount because she knows it is wrong.

Jane has some control in this situation if she knows roughly what the
total of 86 p and 37 p is, and even more control if she can do the addition
quickly and accurately in her head.

EXERCISE 1A Discuss how you would react in each situation and what you need to
know and be able to do so that you feel in control of the situation.

1 Jason needs to get to a cinema by 3 o'clock in the afternoon and it is
vital that he is not late. The journey takes about 35 minutes by bus;
he needs to allow 5 minutes to walk to the bus stop and up to
8 minutes to wait for a bus. He is told to start the journey at
half-past one.

2 Don owes a friend 27 marbles. He gives his friend an unopened bag
with 50 marbles in it and gets back 33 marbles.

3 Sara has scored 246 so far in a game of darts. She needs 301 to win
the game. Ken says 'You need 52 to finish.'

Discussion of these examples should convince you that there are some
situations where it is important to be able to add and subtract numbers
quickly and accurately. There are other situations where it is important
to be able to judge whether a calculation is *roughly* correct. These skills
come with practice.

CONTINUOUS ADDITION OF NUMBERS LESS THAN 100

To add a line of numbers, start at the left-hand side:

$$6 + 4 + 3 + 8 = 21$$

Working in your head
add the first two numbers (10)
then add on the next number (13)
then add on the next number (21)
Check your answer by starting at the other end.

To add a column of numbers, start at the bottom and *working in your head* add up the column:

$$\begin{array}{r} 8 \\ 7 \\ 2 \\ + 5 \\ \hline 22 \end{array} \quad (5 + 2 = 7, 7 + 7 = 14, 14 + 8 = 22)$$

Check your answer by starting at the top and add the column *in your head*.

EXERCISE 1B

Find the value of

____ These are extra questions

1 $2 + 3 + 1 + 4$

2 $1 + 5 + 2 + 3$

3 $5 + 2 + 6 + 1$

4 $3 + 4 + 2 + 6$

5 $5 + 6 + 4 + 2$

<u>6</u> $2 + 1 + 3 + 5$

<u>7</u> $7 + 3 + 8 + 6$

<u>8</u> $5 + 4 + 9 + 1$

<u>9</u> $7 + 3 + 2 + 8$

<u>10</u> $6 + 7 + 5 + 9$

11 $2 + 5 + 4 + 1 + 3$

12 $4 + 8 + 2 + 1 + 2$

13 $6 + 7 + 3 + 5 + 6$

14 $4 + 9 + 2 + 8 + 4$

15 $7 + 3 + 9 + 6 + 8$

<u>16</u> $3 + 2 + 3 + 4 + 1 + 5$

<u>17</u> $4 + 2 + 5 + 6 + 1 + 7$

<u>18</u> $8 + 3 + 9 + 2 + 7 + 3$

<u>19</u> $6 + 9 + 4 + 8 + 7 + 5$

<u>20</u> $4 + 7 + 8 + 6 + 5 + 2$

21	**22**	**23**	**<u>24</u>**	**<u>25</u>**
3	1	4	9	8
7	9	6	7	7
8	5	7	9	6
+6	+2	+3	+8	+9

___ These are
extra questions

26	3	**27**	4	**28**	6	**29**	7	**30**	6
	4		2		5		8		7
	5		3		3		2		3
	1		9		1		1		9
	+8		+3		+4		+8		+7

31	3	**32**	5	**33**	8	**34**	2	**35**	4
	5		7		7		9		8
	2		3		9		5		2
	9		5		2		8		9
	1		4		8		7		9
	+6		+2		+6		+6		+7

36 Find the distance round the edge of this figure.

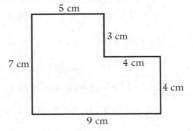

5 cm
3 cm
7 cm
4 cm
4 cm
9 cm

37 a The following sequence of numbers is formed by starting with 8 and then adding 7 to the previous number: 8, 15, 22, 29, ... Write down the next three numbers.

b Write down the first five numbers of a sequence formed by starting with 4 and adding 6 each time.

c Write down the first five numbers of the sequence formed by starting with 6 and adding 9 each time.

38 To score in snooker when there are red balls on the table, a red ball needs to be potted first followed by a colour, then another red followed by another colour, and so on. Each red ball scores 1. The colours, and their scores are:

black 7, pink 6, blue 5, brown 4, green 3, yellow 2.

Peter had three visits to the table at which he potted

a red, black, red, blue, red, pink, red, black

b red, black, red, green, red, black, red, pink

c red, blue, red, yellow, red, pink.

Find Peter's break (i.e. his score) for each session.

This is
a 'challenge'
question

39 In the game of snooker described above, when the last red ball has
been potted, there is a free choice of a coloured ball to follow it.
Once that ball has been potted, it is put back on the table and the
colours then have to be potted in the order yellow, green, brown,
blue, pink, black.
Janice had a turn when there were just two red balls left and she
needed to score 57 to win.
Is it possible to win, if she pots all the balls?

PLACE VALUE

The number one thousand, three hundred and forty-two, can be written
in figures as 1342
The number four thousand, one hundred and twenty-three in figures
is 4123.
The same figures are used but they are in different places.
In 1342, the figure 4 means 4 tens.
In 4123, the figure 4 means 4 thousands.

We can write a number under place headings,

		thousands	hundreds	tens	units
e.g.	4123 can be written as	4	1	2	3
and	3056 can be written as	3	0	5	6

EXERCISE 1C

1 Write in figures the number

  a two hundred and four   **b** ten thousand and thirty

2 Write in words the number

  a 1023   **b** 21 505

3 Write these numbers in order with the smallest first.

  a 207, 89, 1030, 103   **b** 11 020, 90 110, 370 000, 101 010

4 Write down the value of the 5 in the number

  a 152   **b** 2506   **c** 375

5 Make as many different numbers as you can, using each of the figures
2, 6, 7, 0 once in each number.
Arrange the numbers in order with the largest first.

ADDITION OF WHOLE NUMBERS

Sometimes we need to use pencil and paper to add numbers.

To add a column of numbers, start with the units:

$$\begin{array}{r} 83 \\ 291 \\ + \ 702 \\ \hline 1076 \\ \hline \end{array}$$

In the *units* column, $2 + 1 + 3 = 6$ so write 6 in the units column.

In the *tens* column, $0 + 9 + 8 = 17$ tens which is 7 tens and 1 hundred. Write 7 in the tens column and carry the 1 hundred to the hundreds column to be added to what is there already.

In the *hundreds* column, $1 + 7 + 2 = 10$ hundreds which is 0 hundreds and 1 thousand.

EXERCISE 1D

Add up the following numbers in your head.

1	**2**	**3**	**4**	**5**
28	35	22	103	56
$+51$	$+62$	$+43$	$+205$	$+203$

Use pencil and paper to add up these numbers.

___ These are extra questions

6	**7**	**8**	**9**	**10**
101	223	492	259	351
25	317	812	28	1036
$+273$	$+342$	$+735$	$+704$	$+915$

11	**12**	**13**	**14**	**15**
87	9217	3021	93	6943
102	824	84	251	278
56	3216	926	179	5419
$+304$	$+8572$	$+5041$	$+1312$	$+3604$

Find $217 + 85 + 976$

$217 + 85 + 976 = 1278$

It is easier to add numbers when they are in a column.

$$\begin{array}{r} 217 \\ 85 \\ + \ 976 \\ \hline 1278 \\ \hline \end{array}$$

Make sure you line up the units.

Find

16 $28 + 72 + 12$

17 $56 + 10 + 92$

18 $83 + 107 + 52$

19 $256 + 139 + 402$

20 $1026 + 398 + 542$

21 $24 + 83 + 76$

22 $92 + 58 + 27$

23 $52 + 112 + 38$

24 $207 + 394 + 651$

25 $943 + 856 + 984$

26 $599 + 107 + 2058$ **29** $253 + 431 + 1212$

27 $642 + 321 + 4973$ **30** $821 + 903 + 3506$

28 $555 + 921 + 6049$ **31** $727 + 652 + 2716$

32 $92 + 56 + 109 + 324$ **34** $329 + 26 + 73 + 429$

33 $103 + 72 + 58 + 276$ **35** $325 + 293 + 502 + 712$

36
Darts landing in this ring score double

Darts landing in this ring score nothing

Darts landing in these rings score single

Darts landing in this ring score treble

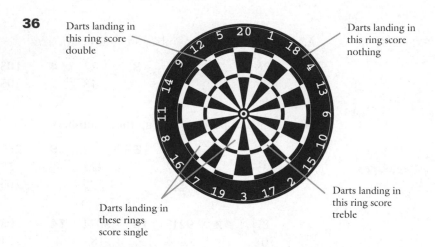

The inner centre ring scores 50 and the outer centre ring scores 25. Find the score on each board.

a   b   c

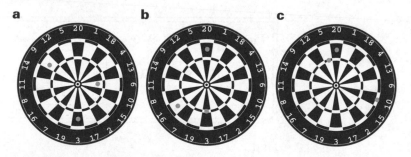

37 When John went to school this morning it took him 4 minutes to walk to the station. He had to wait 12 minutes for the train and the train journey took 26 minutes. He then had an 8 minute walk to his school.
How long did it take John to get to school?

38 Find the sum of one thousand and fifty, four hundred and seven, and three thousand five hundred.

39 A boy decided to save some money by an unusual method. He put
1 p in his money box in the first week, 2 p in the second week, 4 p in
the third week, 8 p in the fourth week, and so on. He gave up after
10 weeks.
Write down how much he put in his money box each week and add
up to find the total that he had saved.
Why do you think he gave up?

40 Complete the number squares using the numbers 1 to 9 once only
in each square to make the total in each row, column and diagonal
the same.

a

2		6
9		
4		8

b

8		
	5	
4		2

c

4	9	
	5	
	1	

**SUBTRACTION
OF WHOLE
NUMBERS**

EXERCISE 1E

Work out the following subtractions in your head.

	1 15	**2** 19	**3** 18	**4** 12	**5** 15
	− 4	− 7	− 4	− 7	− 8

___ These are
extra questions

6 20 − 8   **11** 15 − 2   **16** 11 − 7

7 18 − 3   **12** 12 − 9   **17** 13 − 8

8 17 − 8   **13** 17 − 6   **18** 15 − 9

9 14 − 6   **14** 16 − 8   **19** 20 − 6

10 10 − 4   **15** 19 − 9   **20** 15 − 7

Sometimes we need to use pencil and paper for subtraction. You will
probably have your own method for subtraction. Use it if you
understand it. The worked examples in Exercise 1F show one method.

If you are asked to find the *difference* between two numbers, take the
smaller number from the larger one.

EXERCISE 1F

Find 642 − 316

642 − 316 = 326

$6^3\!4^1 2$
$-3\ 1\ 6$
$\overline{3\ 2\ 6}$

Start with the units column; 2 − 6 is not possible, so take 1 from the number in the tens column. This makes 12 in the units column and leaves 3 in the tens column.

Find 907 − 259

907 − 259 = 648

$^8\!9^9\!0^1 7$
$-2\ 5\ 9$
$\overline{6\ 4\ 8}$

There are no tens here, so take one from the number in the hundreds column. This gives 10 in the tens column, so we can now take 1 from this to give 17 in the units column and leave 9 in the tens column.

Find

___ These are
extra questions

1 526 − 315   **3** 564 − 491   **5** 814 − 344

2 526 − 308   **4** 495 − 369   **6** 592 − 238

7 578 − 291   **10** 1237 − 524   **13** 507 − 499

8 635 − 457   **11** 823 − 568   **14** 3451 − 623

9 602 − 415   **12** 718 − 439   **15** 5267 − 444

16 1027 − 452   **19** 4627 − 3924   **22** 3506 − 3429

17 3927 − 583   **20** 1203 − 527   **23** 7016 − 6824

18 1922 − 398   **21** 4906 − 829   **24** 9342 − 5147

25 The milk bill for last week was 96 p. I paid with a £5 note (£5 is 500 pence).
How much change should I have ?

26 In a school there are 856 children. There are 392 girls.
How many boys are there ?

27 Take two hundred and fifty-one away from three hundred and forty.

28 A shop starts with 750 cans of cola and sells 463.
How many cans are left ?

29 One shop stocks 129 varieties of sweets and another shop stocks 165 varieties.
What is the difference in the number of varieties of sweets stocked by the two shops?

30 Subtract two thousand and sixty-five from eight thousand, five hundred and forty-eight.

31 Find the difference between the value of the 6 in the number 461 and the value of the 3 in the number 307.

32 The sequence of numbers 100, 93, 86, ... is formed by starting with 100 and subtracting 7 each time.
Write down the next five terms of this sequence.

33 Write down the first five terms of a sequence of numbers formed by starting with 94 as the first term and taking away 12 each time.

34 The diagram shows a simple map with distances between villages.

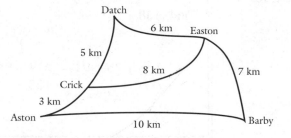

How far is it from Aston to Barby

a via Datch **b** via Crick but not Datch?

35 The road I live in has 97 houses in it. The road my friend lives in has 49 houses in it.
How many more houses are there in my road than in my friend's road?

36 Ben Nevis is 1343 m high and is the highest mountain in Great Britain.
Mount Everest is 8843 m high.
How much higher than Ben Nevis is Mount Everest?

This is
a 'challenge'
question

37 Peter needs a total score of 301 to win a game of darts. On his first turn he scores 58, on his second turn he scores 75. How many does he still have to score to win?

38 Find the missing digit; it is marked with □.

a $27 + 38 = \square 5$ **d** $5\square + 29 = 83$

b $34 + 5\square = 89$ **e** $\square 4 + 57 = 81$

c $128 + \square 59 = 1087$ **f** $1\square 7 + 239 = 416$

MIXED ADDITION AND SUBTRACTION

Some problems, like question **37** in the last exercise, can involve a mixture of addition and subtraction.

It is the sign *in front of* a number that tells you what to do with that number. For example $128 - 56 + 92$ means '128 take away 56 and add on 92'. This can be done in any order so we could add on 92 and then take away 56, i.e.

$$128 - 56 + 92 = 220 - 56$$
$$= 164$$

EXERCISE 1G

Find $138 + 76 - 94$

$$138 + 76 - 94 = 214 - 94$$
$$= 120$$

$$\begin{array}{r} 138 \\ + 76 \\ \hline 214 \end{array}$$

$$\begin{array}{r} 214 \\ - 94 \\ \hline 120 \end{array}$$

Find $56 - 72 + 39 - 14$

$$56 - 72 + 39 - 14 = 95 - 72 - 14$$
$$= 23 - 14$$
$$= 9$$

$$\begin{array}{r} 56 \\ + 39 \\ \hline 95 \end{array}$$

$$\begin{array}{r} 95 \\ - 72 \\ \hline 23 \end{array}$$

Find

These are
extra questions

1 $25 - 6 + 7 - 9$ **5** $46 - 12 + 3 - 9$

2 $14 + 2 - 8 - 3$ **6** $27 + 6 - 11 - 9$

3 $7 - 4 + 5 - 6$ **7** $2 + 13 - 7 + 3 - 8$

4 $23 - 2 + 4 + 5$ **8** $17 + 4 - 9 - 3 - 5$

___ These are extra questions

9 $17 - 9 + 11 - 19$

10 $36 - 24 + 62 - 49$

11 $51 - 27 - 38 + 14$

12 $124 + 51 - 78 - 14$

13 $91 - 50 + 36 - 27$

14 $105 + 23 - 78 - 50$

15 $73 - 42 - 19 + 27$

16 $361 - 200 + 15 - 81$

17 $213 - 307 + 198 - 31$

18 $29 + 108 - 210 + 93$

19 $493 - 1000 + 751 - 140$

20 $612 - 318 + 219 + 84$

21 $95 - 161 + 75 + 10$

22 $952 - 1010 - 251 + 438$

23 A boy buys a comic costing 22 p and a pencil costing 18 p. He pays with a 50 p piece.
How much change does he get?

24 Find the sum of eighty-six and fifty-four and then take away sixty-eight.

25 I have a piece of string 200 cm long. I cut off two pieces, one of length 86 cm and one of length 34 cm. How long is the piece of string that I have left?

26 On Monday 1000 fish fingers were cooked in the school kitchen. At the first sitting 384 fish fingers were served. At the second sitting 298 fish fingers were served.
How many were left?

27 Find the difference between one hundred and ninety, and eighty-three. Then add on thirty-seven.

28 A greengrocer has 38 lb of carrots when he opens on Monday morning. During the day he gets a delivery of 60 lb of carrots and sells 29 lb of carrots.
How many pounds of carrots are left when he closes for the day?

29 A boy has 30 marbles in his pocket when he goes to school on Monday morning. At first playtime he wins 6 marbles. At second playtime he loses 15 marbles. At third playtime he loses 4 marbles. How many marbles does he now have?

___ These are extra questions

30 Sarah gets £2.00 pocket money on Saturday. On Monday she spends 84 p. On Tuesday she is given 50 p for doing a special job at home. On Thursday she spends 47 p.
How much money does she have left?

31 The sequence of numbers

$$7, 16, 11, 20, 15, \ldots$$

is made by starting with 7, adding 9 to get the next number, subtracting 5 to get the next number, adding 9, subtracting 5, and so on.

a Write down the next 4 numbers in the sequence.

b Starting with 7, write down the third, fifth, seventh numbers, and so on, to make another sequence.
What do you think the rule is for generating this sequence?

32 The weekly profit made in a school tuck shop is found by taking away the cost of the items bought for the shop that week from the total takings for the week.
(The pupils run the shop and do not get paid!)
Find the profit when £180 was spent on items for the shop and the takings were £249.

33 In question **32** a general instruction was given for finding the weekly profit and then the profit was found for a particular case. Here is the general instruction for obtaining a sequence of numbers:

The next number is obtained by adding the previous two numbers.

a The first number is 3 and the second number is 5.
Write down the next six numbers.

b The first number is 4 and the second number is 2.
Write down the next six numbers.

___ This is a 'challenge' question

34 Write down a general instruction for obtaining the sequence of numbers

a 1, 3, 5, 7, 9, ...   **b** 50, 42, 34, 26, 18, ...

APPROXIMATION

If you were told that a games machine priced at £147 and a pack of accessories priced at £39 will cost you a total of £205, would you know quickly that the total quoted was wrong?

There are situations when we cannot do a calculation in our heads and when there is not time or opportunity to use pencil and paper. If a calculator is available, we can use that, but we sometimes make mistakes when using them. It is important to know if a result is roughly right. By simplifying numbers it is possible to get a rough answer in our heads.

One way to simplify numbers is to make them into the nearest number of tens. For example

$$127 \text{ is roughly 13 tens, or 130}$$

and

$$123 \text{ is roughly 12 tens, or 120}$$

We say that 127 is *rounded up* to 130 and 123 is *rounded down* to 120. In mathematics we say that 127 is approximately equal to 13 tens.

We use the symbol $\approx$ to mean 'is approximately equal to'.
We would write

$$127 \approx 13 \text{ tens}$$

and

$$123 \approx 12 \text{ tens}$$

When a number is half way between tens we always round up. We say

$$125 \approx 13 \text{ tens}$$

EXERCISE 1H

Write 56 as an approximate number of tens.

$$56 \approx 6 \text{ tens}$$

Write each of the following numbers as an approximate number of tens.

1 84	**3** 46	**5** 8	**7** 228	**9** 73
2 151	**4** 632	**6** 37	**8** 155	**10** 4

Write 1278 as an approximate number of hundreds.

$$1278 \approx 13 \text{ hundreds}$$

Write each of the following numbers as an approximate number of hundreds.

11 830	**13** 780	**15** 1350	**17** 1560	**19** 972
12 256	**14** 1221	**16** 450	**18** 3780	**20** 1965

> By writing each number correct to the nearest number of tens find
> an approximate answer for $196 + 58 - 84$.
>
> $$196 \approx 20 \text{ tens}$$
> $$58 \approx 6 \text{ tens}$$
> $$84 \approx 8 \text{ tens}$$
> Therefore $196 + 58 - 84 \approx 20 \text{ tens} + 6 \text{ tens} - 8 \text{ tens}$
> $$\approx 18 \text{ tens} = 180$$

By writing each number correct to the nearest number of tens find an
approximate answer for

___ These are
extra questions

21 $344 - 87$

22 $95 - 39$

23 $258 - 49$

24 $153 + 181$

25 $89 - 51$

26 $258 + 108$

27 $391 - 127$

28 $832 - 55$

29 $83 + 27 - 52$

30 $76 - 31 - 29$

31 $137 - 56 + 82$

32 $295 + 304 - 451$

33 $49 - 25 + 18$

34 $68 + 143 + 73$

35 $153 + 19 + 57$

36 $250 + 31 - 121$

37 $127 + 56 + 82 + 95$

38 $73 + 21 + 37 + 46 + 29$

39 $83 + 64 + 95 + 51$

40 $63 + 29 + 40 + 37 + 81$

Now use your calculator to find the exact answers in questions **21** to **40**.
Remember to look at your rough answer to check that your calculator
answer is probably right.

There are four answers given for each calculation. Three of them are
obviously wrong. Without working them out, write down the letter of
the answer that *might* be correct.

41 $257 - 32$:   A 25   B 289   C 225   D 125

42 $749 + 412$:   A 337   B 1161   C 2550   D 961

43 $290 - 181$:   A 9   B 89   C 109   D 471

44 $682 + 798 - 56$:   A 2784   B 424   C 1424   D 882

45 $1278 - 569$:   A 209   B 709   C 1847   D 967

**NUMBER
PUVVLES**

1

Each disc shows three digits. Remove one digit from one of the discs and place it on another disc so that the digits on each of the three discs have the same total.

2 Follow these instructions.

Step 1 Write down any three-figure number in which the number of hundreds differs by at least two from the number of units, e.g. 419 or 236 or 973 but not 707 or 514.

Step 2 Now write down the digits in reverse order, e.g. 419 becomes 914.

Step 3 Subtract the smaller number from the larger number, e.g. 914 − 419 = 495.

Step 4 Add this number to its reverse, i.e. add 495 to 594.
The result in this case is 1089.

Now try these four steps with any number of your own choice. Repeat the instructions six times for six different numbers, but do remember that the number of hundreds must differ by at least two from the number of units. What do you notice?

3 Solve the following cross-number puzzle.

Across   **1** 73 − 31
             4 249 + 167
             6 700 − 565
             7 231 − 158
Down   **2** 77 + 166
             3 222 − 136
             5 78 + 79
             6 52 + 106 − 139

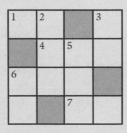

4 Use the digits 1, 2, 3, 4, 5, 6, 7, 8 and 9 in order, using each digit once, to make four numbers separated by either a + sign or a − sign so that the answer is 100.

5 Copy this diagram on to a sheet of paper.

	7		5	
9	2		3	4
	6	1	8	

Cut it into three pieces and fit them together to form a magic square.

INVESTIGATIONS

1 Sonia makes up a pattern starting with 3 and 4. To get the next number in the pattern she adds the previous two numbers together. If the answer is more than 10 she writes down only the number of units.

Her pattern is 3, 4, 7, 1, 8, 9, 7, ...

Write down the next ten numbers in this pattern.

Does the pattern repeat itself?

If so, how many numbers are there before it starts to repeat?

Now start with 4 and 3 and see what happens. (You need to keep going for a long time!)

Investigate some other pairs of numbers.

2 Write down any three-digit number, e.g. 287

Arrange the digits in order of size; once with the largest digit first and once with the smallest digit first, i.e. 872, 278

Now find the difference between these two numbers (take the smaller number from the larger number),

i.e. $872 - 278 = 594$

Repeat the process for your answer,

i.e. $954 - 459 = 495$

Repeating the process again,

i.e. $954 - 459 = 495$

These rules give us a chain of numbers. In this example we have

$$287 \rightarrow 594 \rightarrow 495 \rightarrow 495$$

Form a similar chain for a three-figure number of your own choice. Try a few more.

What do you notice?

3 A number that reads the same forwards and backwards, e.g. 14 241 is called a palindrome.

There is a conjecture (i.e. it has not been proved) that if we take *any* number, reverse the digits and add the numbers together, then do the same with the result, and so on, we will end up with a palindrome,

e.g. starting with 251,
$$251 + 152 = 403$$
$$403 + 304 = 707 \quad \text{which is a palindrome.}$$

a Can you find some two digit numbers for which the palindrome appears after the first sum ?

b If two palindromes are added together, is the result always a palindrome ?

4 The Romans did not use symbols for numbers, but used letters of the alphabet. For example, in Roman numerals, X is used for ten, V for five; XV means 'ten and five', i.e. 15.

The numbers one to six are written I, II, 111, IV, V, VI.

a Use reference material to find out what the following Roman numbers are.

XIV, CLX, MLII

b Write the following numbers in Roman numerals.

25, 152, 1854, 2002

c Find LXII − XXIV.

MULTIPLICATION AND DIVISION WITH WHOLE NUMBERS

Jake had to collect 27 pence from each of the 8 members of his youth group. He didn't keep the money separate from his own so when he needed to hand the total to his group leader, he used his calculator to work out 27×8.

The display read 296, so he handed over £2.96. This is 80 p more than he should have given.

Jake clearly made a mistake when he used his calculator, but he didn't know that he had. This meant that he was quite unaware that he was losing 80 p.

If Jake could

- work out *in his head* that
  27×8 is a bit less than 30×8 which is 240,
  he would know that £2.96 was too much

- work out, in his head or on paper, that
  $27 \times 8 = 216$,
  he would not have relied on a calculator.

EXERCISE 2A

Discuss what you need to know, and be able to do, so that you have some control in these situations.

1   Ali has to order coaches to take 218 children on a school outing. Each coach has 45 seats available. The coach company says it will send 5 coaches.

2   John goes into a shop with 35 p to buy some sweets. He chooses 3 sweets costing 5 p each, 4 sweets costing 3 p each and a lolly costing 12 p. At the cash desk he finds that he does not have enough money to pay for all these things, and has to put some back.

3   Anna needs to buy a ladder that will extend to at least 600 cm in length. The salesman recommends a ladder with three sections, each 250 cm long. When the ladder is fully opened out, there is an overlap of 30 cm between the sections.

USING THE MULTIPLICATION FACTS

The basic knowledge that underpins all multiplication and division is knowing the multiplication tables. By 'know' we mean instant recall of the product of any pair of numbers from the list 1, 2, 3, 4, 5, 6, 7, 8, 9.

For example 69×4 *can* be found by adding

$$69 + 69 + 69 + 69$$

but, if you know them, it is quicker to use the multiplications facts. You will probably have your own methods for multiplying, but here is one way of calculating 69×4.

$$69 \times 4 = 280 - 4$$
$$= 276$$

$69 = 70 - 1$, so 69×4 can be found from $(70 \times 4) - (1 \times 4)$

If you are confident in your ability to multiply and add, you can do this in your head. If you need to use paper, it helps to set the calculation out like this:

$$
\begin{array}{r}
69 \\
\times 4 \\
\hline
276 \\
{\scriptstyle 3}
\end{array}
$$

9 units $\times$ 4 = 36 units; write 6 under the units column and 'carry' the 3 by writing it underneath the line in the tens column.
6 tens $\times$ 4 = 24 tens; add on the 3 tens carried to give 27 tens, i.e. 2 hundreds and 7 tens.

The next few exercises help you practice the multiplication skills you need.

1 a Copy and complete the following multiplication table. Time yourself; you should be able to fill in the blank spaces in less than 2 minutes.

×	3	7	5	2	4	6	9	8	1
4									
1									
6									
2									
9									
3									
5									
8									
7									

b If you filled in this table by working across the rows, repeat the exercise by working down the columns (or vice-versa).

c Now choose the squares you fill in at random.

Find 24×8

> If you can do this in your head, write the answer down.
> If you need to use paper, most people find it easier to set the multiplication in a column.

$$\begin{array}{r} 24 \\ \times\ \ 8 \\ \hline 192 \\ \scriptstyle 3 \end{array}$$

$24 \times 8 = 192$

Do these in your head if you can. If you cannot, use pencil and paper. Do not use a calculator.

Find

2 23×2		**5** 76×4		**8** 25×4		**11** 83×5
3 42×3		**6** 58×5		**9** 16×9		**12** 47×3
4 13×8		**7** 31×3		**10** 72×2		**13** 54×6

14 21×6		**16** 73×4		**18** 67×8		**20** 8×21
15 84×7		**17** 2×81		**19** 73×9		**21** 7×32

22 152×4 **25** 194×2 **28** 953×3 **31** 312×7

23 307×8 **26** 221×9 **29** 204×8 **32** 142×6

24 256×3 **27** 211×4 **30** 876×3 **33** 513×5

34 Munchy bars cost 32 p each.
How much do 8 of these bars cost?

35 A school day is 7 hours long.
How many minutes is this?

36 One jar of jam weighs 516 grams.
What is the weight of 7 jars of jam?

37 Carol can walk up a flight of steps at the rate of 30 steps a minute.
It takes her 3 minutes to reach the top.
How many steps are there?

38 Peter bought 6 pads of file paper priced at £1.29 each.
He was asked for £6.50. This is not correct; without working
out 129×6, decide if it is too much or too little.
Explain how you decided.

39 A car travelling at 50 miles an hour took 3 hours to travel from
London to Cardiff.
How many miles did the car travel?

40 a The sequence of numbers 3, 12, 48, ... is formed by starting
with 3 and then multiplying the last number by 4.
Write down the next two numbers in the sequence.

    b Write down the first five numbers of the sequence formed by
starting with 7 and multiplying by 2 each time.

    c Write down, in words, the rule for generating this sequence:

$$5, 25, 125, 625, 3125, \ldots$$

41 Write down the missing digit in the following products.

    a $2\square \times 5 = 135$ **b** $16 \times \square = 48$ **c** $\square 4 \times 7 = 168$

MULTIPLICATION BY 10, 100, 1000, …

When 85 is multiplied by 10, the 5 units become 5 tens and the 8 tens become 8 hundreds. So

$$85 \times 10 = 850$$

When 85 is multiplied by 100 the 5 units become 5 hundreds and the 8 tens become 8 thousands. Thus

$$85 \times 100 = 8500$$

When 85 is multiplied by 20 this is the same as $85 \times 2 \times 10$.

So
$$85 \times 20 = 85 \times 2 \times 10$$
$$= 170 \times 10 = 1700$$

In the same way

$$27 \times 4000 = 27 \times 4 \times 1000 = 108 \times 1000 = 108\,000$$

EXERCISE 2C

Find

1 27×10

2 82×100

3 36×10

4 108×10

5 256×1000

6 392×10

Find 42×900

$$42 \times 900 = 42 \times 9 \times 100$$
$$= 378 \times 100$$
$$= 37\,800$$

We find 42×9 first and then multiply by 100.

Find

7 21×20

8 42×300

9 31×40

10 39×200

11 52×50

12 73×400

13 58×60

14 221×30

15 121×700

16 73×2000

17 39×900

18 157×60

19 295×80

20 88×70

21 350×200

22 609×80

23 270×200

24 556×70

25 81×3000

26 390×90

27 107×400

28 240×80

29 100×88

30 200×95

31 856×70

32 How much would you have to pay for 200 tickets costing £14 each?

33 What is the total number of cans in 400 cartons when each carton contains 48 cans?

34 Which of these answers are obviously wrong for $253 \times 2000 =$

A 50 600 **B** 506 000 **C** 5 060 000

35 There are 250 nails in each packet. Jon needed to know the number of nails in 400 packets..
Which of these answers are obviously wrong?

A 100 000 **B** 25 000 **C** 10 000

36 Jane was asked how many balloons there were in 200 packets, when each packet had 50 balloons in it. She said 1000 which is wrong. She tried to justify her answer by saying '5 times 2 is ten, and then add two noughts for the 200.'
What is wrong with her reasoning?

LONG MULTIPLICATION

To multiply 84×26 we use the fact that
$84 \times 26 = 84 \times 6 + 84 \times 20$
This can be set out as

$$
\begin{array}{r}
84 \\
\times 26 \\
\hline
504 \quad (84 \times 6) \\
+1680 \quad (84 \times 20) \\
\hline
2184
\end{array}
$$

EXERCISE 2D

Find 2813×402

$$
\begin{array}{r}
2813 \\
\times \quad 402 \\
\hline
5626 \\
+1125200 \\
\hline
1130826
\end{array}
\quad
\begin{array}{l}
(2813 \times 2) \\
(2813 \times 400) \\
\end{array}
$$

$2813 \times 402 = 1\,130\,826$

Find

1 32×21 **6** 38×41 **11** 241×32

2 43×13 **7** 107×26 **12** 153×262

3 86×15 **8** 53×82 **13** 433×921

4 27×21 **9** 74×106 **14** 1251×28

5 34×42 **10** 36×89 **15** 3421×33

16 512×210 **21** 2004×43 **26** 385×95

17 487×82 **22** 584×97 **27** 750×450

18 724×98 **23** 187×906 **28** 605×750

19 146×259 **24** 270×709 **29** 1008×908

20 805×703 **25** 3060×470 **30** 1500×802

USING A CALCULATOR FOR LONG MULTIPLICATION

Calculators save a lot of time when used for long multiplication. You do, however, need to be able to estimate the size of answer you expect, as a check on your use of the calculator.

One way to get a rough answer is to round

a number between 10 and 100 to the nearest number of tens
a number between 100 and 1000 to the nearest number of hundreds
a number between 1000 and 10 000 to the nearest number of thousands

and so on.

For example $\quad 512 \times 78 \approx 500 \times 80 = 40\,000$

and $\quad\quad\quad 2752 \times 185 \approx 3000 \times 200 = 600\,000$

EXERCISE 2E Estimate

1 79×34 **6** 59×18 **11** 159×93

2 29×27 **7** 23×55 **12** 82×309

3 84×36 **8** 62×57 **13** 281×158

4 45×32 **9** 136×29 **14** 631×479

5 87×124 **10** 52×281 **15** 273×784

Estimate 2581×39 and then use a calculator to find the exact value.

$2581 \times 39 \approx 3000 \times 40 = 120\,000$ (estimate)

$2581 \times 39 = 100\,659$ (calculator)

> The estimate and calculator answer agree quite well, so the calculator answer is probably correct.

Estimate the answer and then use your calculator to work out the following.

16 258×947 **21** 78×91 **26** 52×821

17 29×384 **22** 625×14 **27** 89×483

18 182×56 **23** 33×982 **28** 481×97

19 37×925 **24** 2501×12 **29** 608×953

20 782×24 **25** 87×76 **30** 4897×61

31 69×78 **34** 463×87 **37** 37×634

32 47×853 **35** 271×82 **38** 541×428

33 94×552 **36** 753×749 **39** 798×583

40 There are three answers given for each calculation. Two of them are obviously wrong.
Write down the letter of the answer that *might* be correct.

  a 521×36: **A** 1876   **B** 11 886   **C** 18 756

  b 63×95: **A** 6005   **B** 585   **C** 10 000

41 Multiply three hundred and fifty-six by twenty-three.

42 One jar of marmalade weighs 454 grams.
Find the weight of 124 jars.

43 Find the value of one hundred and fifty multiplied by itself.

44 A car park has 34 rows and each row has 42 parking spaces.
How many cars can be parked?

45 A supermarket takes delivery of 54 crates of soft drink cans. Each crate contains 48 cans.
How many cans are delivered?

46 A light bulb was tested by being left on non-stop. It failed after 28 days exactly.
For how many hours was it working?

47 Fill in the missing digits in these calculations.

  a $2\square \times 42 = 966$

  b $125 \times \square 6 = 7000$

  c $157 \times 3\square = 5495$

WHEN NOT TO USE A CALCULATOR

Some people use their calculators for all number calculations. This is silly for two reasons.

- It is quicker to use your head for simple numbers. (The next exercise should convince you that this is true.)

- If you rely on your calculator for all number work, you will forget the number facts. This means you will lose the ability to judge when an answer obtained from a calculator is obviously wrong.

EXERCISE 2F Work in pairs. Copy the square once each. One of you should fill it in, working in your head, while being timed by your partner. Now reverse roles but this time use a calculator to find each number.

1

×	2	3	4
2			
4			
6			

3

×	5	10	3
3			
12			
5			

2

+	5	2	7
9			
8			
4			

4

+	2	4	6
7			
3			
2			

DIVISION OF WHOLE NUMBERS

To find out how many eight-seater mini-buses are needed to take 36 children on an outing, we need to find how many eights there are in 36. $36 \div 8$ means 'how many eights are there in 36?'.

We could find out by repeatedly taking 8 away from 36:

$$36 - 8 = 28$$
$$28 - 8 = 20$$
$$20 - 8 = 12$$
$$12 - 8 = 4 \qquad \text{So there are 4 eights in 36 with 4 left over.}$$

Therefore $36 \div 8 = 4$, remainder 4.

However, a quicker way uses the multiplication facts.
We know that $32 = 4 \times 8$
therefore $36 \div 8 = 4$, remainder 4

Now we know that we need five mini-buses.

Larger numbers can be divided in a similar way.

To find $534 \div 3$ start with the hundreds:

$$5\,(\text{hundreds}) \div 3 = 1\,(\text{hundred}), \quad \text{remainder } 2\,(\text{hundreds})$$

Take the remainder, 2 (hundreds), and add to the tens:

$$23\,(\text{tens}) \div 3 = 7\,(\text{tens}), \quad \text{remainder } 2\,(\text{tens})$$

Take the remainder, 2 (tens), and add to the units:

$$24\,(\text{units}) \div 3 = 8 \text{ units}$$

Therefore $\qquad 534 \div 3 = 178$

This can be set out as $\quad 3\overline{)5^2 3^2 4}$ with 178 above.

EXERCISE 2G

> Calculate $4509 \div 5$, giving the remainder.
>
> $4509 \div 5 = 901, \text{r}\,4$ $\qquad\qquad 5\overline{)4509} = 901\,\text{r}\,4$

Calculate, giving the remainder when there is one.

1 $87 \div 3$	**6** $97 \div 2$	**11** $78 \div 8$
2 $56 \div 4$	**7** $73 \div 5$	**12** $85 \div 7$
3 $36 \div 6$	**8** $83 \div 4$	**13** $39 \div 3$
4 $57 \div 3$	**9** $69 \div 3$	**14** $21 \div 9$
5 $72 \div 4$	**10** $82 \div 6$	**15** $78 \div 6$
16 $54 \div 2$	**20** $855 \div 5$	**24** $294 \div 9$
17 $639 \div 3$	**21** $693 \div 3$	**25** $570 \div 7$
18 $548 \div 2$	**22** $721 \div 7$	**26** $680 \div 8$
19 $605 \div 3$	**23** $358 \div 5$	**27** $731 \div 6$

28 $3501 \div 3$	**32** $6405 \div 6$	**36** $1788 \div 9$
29 $1763 \div 4$	**33** $7399 \div 5$	**37** $1098 \div 6$
30 $4829 \div 2$	**34** $8772 \div 4$	**38** $2481 \div 7$
31 $1758 \div 5$	**35** $9712 \div 8$	**39** $6910 \div 4$

40 How many apples costing 8 p each can I buy for 65 p?

41 Cara saves the same amount each week. After 7 weeks she has saved £21.
How much does she save each week?

42 A greengrocer bought a sack of potatoes weighing 50 kg. He divided the potatoes into bags, so that each bag held 3 kg of potatoes.
How many complete bags of potatoes did he get from his sack?

43 Three children are given 60 p to split equally among them.
How much does each child get?

44 Jim is paid £205 for a five day working week.
How much is he paid for each day?

45 How many times can 5 be taken away from 132?

46 How can you tell, without doing long division or using a calculator, that 153 918 is not divisible exactly by 4?

DIVISION BY 10, 100, 1000, ...

$812 \div 10$ means 'how many tens are there in 812?'.
There are 81 tens in 810 so

$$812 \div 10 = 81, \qquad \text{remainder 2}$$

$2578 \div 100$ means 'how many hundreds are there in 2578?'.
There are 25 hundreds in 2500 so

$$2578 \div 100 = 25, \qquad \text{remainder 78}$$

EXERCISE 2H

Calculate and give the remainder.

1 $256 \div 10$	**5** $4910 \div 1000$	**9** $9426 \div 1000$
2 $87 \div 10$	**6** $57 \div 10$	**10** $8512 \div 100$
3 $196 \div 100$	**7** $186 \div 10$	**11** $3077 \div 100$
4 $2783 \div 100$	**8** $2781 \div 10$	**12** $5704 \div 1000$

LONG DIVISION

To find $2678 \div 21$ we can use the same method as for short division but we write the working down more fully.

$$\begin{array}{r} 127 \\ 21\overline{)2678} \\ 21\downarrow \\ \overline{57} \\ 42\downarrow \\ \overline{158} \\ 147 \\ \overline{11} \end{array}$$

There is 1 twenty-one in 26, r 5 (hundreds).

There are 2 twenty-ones in 57, r 15 (tens).

There are 7 twenty-ones in 158, r 11 (units).

So $2678 \div 21 = 127$, r 11.

EXERCISE 2I

Find $2606 \div 25$, giving the remainder.

$$\begin{array}{r} 104 \\ 25\overline{)2606} \\ 25 \\ \overline{106} \\ 100 \\ \overline{6} \end{array}$$

$2606 \div 25 = 104$, r 6

Calculate and give the remainder.

If you use your calculator to check your answers, it will give the whole number part of the answer but it will not give the remainder as a whole number.

1 $254 \div 20$ **5** $394 \div 19$ **9** $389 \div 23$

2 $685 \div 13$ **6** $267 \div 32$ **10** $298 \div 14$

3 $739 \div 41$ **7** $875 \div 25$ **11** $433 \div 15$

4 $862 \div 25$ **8** $269 \div 16$ **12** $614 \div 27$

13 $2804 \div 13$ **17** $2943 \div 23$ **21** $7514 \div 34$

14 $7315 \div 21$ **18** $2694 \div 31$ **22** $5829 \div 43$

15 $8392 \div 34$ **19** $1875 \div 25$ **23** $6372 \div 27$

16 $6841 \div 15$ **20** $3621 \div 30$ **24** $8261 \div 38$

25 $7315 \div 24$ **30** $2694 \div 30$ **35** $8200 \div 250$

26 $8602 \div 15$ **31** $8013 \div 40$ **36** $3606 \div 300$

27 $3004 \div 31$ **32** $829 \div 106$ **37** $8491 \div 150$

28 $1608 \div 25$ **33** $5241 \div 201$ **38** $7625 \div 302$

29 $7092 \div 35$ **34** $3689 \div 151$ **39** $1092 \div 206$

40 4000 apples are packed into boxes, each box holding 75 apples. How many boxes are required?

41 A vegetable plot is 1000 cm long. Cabbages are planted in a row down the length of the plot.
If the cabbages are planted 30 cm apart and the first cabbage is planted 5 cm from the end, how many cabbages can be planted in one row?

42 A class is told to work out the odd-numbered questions in an exercise containing 29 questions.
How many questions do they have to do?

43 Find the missing digits in the following calculations.

     a $256 \div 1\square = 16$

     b $86\square \div 24 = 36$

     c $105\square \div 25 = 42, \text{r } 2$

MIXED OPERATIONS OF $+, -, \times, \div$

When a calculation involves a mixture of the operations $+, \ -, \ \times, \ \div$ we always do

> multiplication and division first

For example

$2 \times 4 + 3 \times 6 = 8 + 18$ (Do the multiplication first, i.e. 2×4 and 3×6 first.)

$\qquad\qquad\qquad\quad = 26$

EXERCISE 2J

Find $2 + 3 \times 6 - 8 \div 2$

$$2 + 3 \times 6 - 8 \div 2 = 2 + 18 - 4$$
$$= 16$$

> Work out 3×6 and $8 \div 2$ first.

Find $5 - 10 \times 2 \div 5 + 3$

$$5 - 10 \times 2 \div 5 + 3 = 5 - 20 \div 5 + 3$$
$$= 5 - 4 + 3$$
$$= 4$$

> Do the multiplication first. Now do the division.

Find

1 $2 + 4 \times 6 - 8$

2 $24 \div 8 - 3$

3 $6 + 3 \times 2$

4 $7 \times 2 + 6 - 1$

5 $18 \div 3 - 3 \times 2$

6 $7 + 4 - 3 \times 2$

7 $8 \div 2 + 6 \times 3$

8 $14 \times 2 \div 7 - 3 + 6$

9 $6 - 2 \times 3 + 7$

10 $5 + 4 \times 3 + 8 \div 2$

11 $7 + 3 \times 2 - 8 \div 2$

12 $5 \times 4 \div 2 + 7 \times 2$

13 $6 \times 3 - 8 \times 2$

14 $9 \div 3 + 12 \div 6$

15 $12 \div 3 - 15 \div 5$

16 $9 + 3 - 6 \div 2 + 1$

17 $6 - 3 \times 2 + 9 \div 3$

18 $7 + 2 \times 4 - 8 \div 4$

19 $7 \times 2 + 8 \times 3 - 2 \times 6$

20 $5 \times 3 \times 2 - 4 \times 3 \div 2$

21 $19 + 3 \times 2 - 8 \div 2$

22 $7 \times 2 - 3 + 6 \div 2$

23 $8 + 3 \times 2 - 4 \div 2$

24 $7 \times 2 - 4 \div 2 + 1$

25 $6 + 8 \div 4 + 6 \times 3 \div 2$

26 $5 \times 3 \times 4 \div 12 + 6 - 2$

27 $5 + 6 \times 2 - 8 \div 2 + 9 \div 3$

28 $7 - 9 \div 3 + 6 \times 2 - 4 \div 2$

29 $9 \div 3 - 2 + 1 + 6 \times 2$

30 $4 \times 2 - 6 \div 3 + 3 \times 2 \times 4$

In the greengrocer's I bought 3 oranges that cost 12 p each and one cabbage that cost 36 p. I paid with a £1 coin.
How much change did I get?

$$
\begin{aligned}
\text{Cost of the oranges} &= 36\,\text{p} \\
\text{Cost of the cabbage} &= 36\,\text{p} \\
\text{Total cost} &= 72\,\text{p} \\
\text{Change from} &= 100\,\text{p} - 72\,\text{p} \\
&= 28\,\text{p}
\end{aligned}
$$

31 I bought 5 oranges that cost 10 p each and 2 lemons that cost 9 p each. How much did I spend?

32 Three children went into a sweet shop. The first child bought three sweets costing 2 p each, the second child bought three sweets costing 1 p each and the third child bought three sweets costing 3 p each. How much money did they spend altogether?

33 A club started the year with 82 members. During the year 36 people left and 28 people joined.
How many people belonged to the club at the end of the year?

34 One money box has five 5 p pieces and four 10 p pieces in it. Another money box has six 10 p pieces and ten 2 p pieces in it.

a What is the total sum of money in the two money boxes?

b Which box has more money in it? How much more is there in this box than the other box?

35 At a school election one candidate got 26 votes, and the other candidate got 35 votes. 10 voting papers were spoiled and 5 pupils did not vote.
How many children altogether could have voted?

36 An extension ladder is made of three separate parts, each 300 cm long. When it is fully extended, there is an overlap of 30 cm at each junction.
How long is the extended ladder?

37 Jane, Sarah and Claire come to school with 20 p each. Jane owes Sarah 10 p and she also owes Claire 5 p. Sarah owes Jane 4 p and she also owes Claire 8 p.
When all their debts are settled, how much money does each girl have?

38 The total number of children in the first year of a school is 500. There are 50 more girls than boys. How many of each are there?

39 In a book of street plans of a town, the street plans start on page 6 and end on page 72. How many pages of street plans are there?

40 Sam said that $25 + 5 \times 5$ was equal to 150. Explain his mistake and give the correct answer.

41 A mountaineer starts from a point which is 150 m above sea level. He climbs 200 m and then descends 50 m before climbing another 300 m. How far is he now above sea level?

42 The calculation for the worked example above question **31** can be set out using the numbers given, as $100 - 3 \times 12 - 36$. Set out the calculations for questions **31** to **36** in the same way using only the numbers given in the questions.

43 My great-grandmother died in 1894, aged 62. In which year was she born?

44 An oak tree was planted in the year in which Lord Swell was born. He died in 1940, aged 80. How old was the oak tree in 1984?

45 A bus leaves the bus station at **9.30** a.m. It reaches the Town Hall at **9.40** a.m. and gets to the railway station at **9.52** a.m.

 a How long does it take to go from the Town Hall to the railway station?

 b Which piece of information given in the question is not needed to answer part **a**?

MIXED OPERATIONS USING A CALCULATOR

A scientific calculator obeys the rule 'multiplication and division before addition and subtraction'.

When $2 + 3 \times 6$ is keyed in as `2` `+` `3` `x` `6`, the calculator will work out 3×6 first.

If we need to do addition or subtraction first, we can use brackets to show this, i.e. $(2 + 3) \times 6$ means 'add 2 and 3 first'.

We can use the bracket keys on a calculator to find this in one operation,

i.e. press `(` `2` `+` `3` `)` `x` `6` `=`

We can also do this on a calculator by pressing the `=` key after entering $2 + 3$, then we carry on with $\times 6$, i.e. press

`2` `+` `3` `=` `x` `6` `=`

EXERCISE 2K

Use your calculator to find $2578 - (36 + 24) \times 13$.

Estimate: $2578 - (36 + 24) \times 13 \approx 3000 - (40 + 20) \times 10$
$= 3000 - 600 = 2400$

$2578 - (36 + 24) \times 13 = 1798$

Notice that 3000 is a big overestimate for 2578 and 600 is a large underestimate for $(36 + 24) \times 13$. This means that 2400 is a *very rough* estimate, but if the calculator result is over 10 000 or under 1000 we should be suspicious.

Use your calculator to find the answers. Do not forget to estimate the answer first.

1 $25 + 52 \times 26$

2 $279 \times (32 + 27)$

3 $(25 + 52) \times 26$

4 $398 \times 24 - 2560$

5 $579 + (46 - 37) \times 16$

6 $2965 \times 36 - 293 \times 178$

7 $36 - 216 \div (21 - 33)$

8 $4088 \div 73 - 29$

9 $27 + (1201 + 17) \div 58$

10 $1380 - (120 \times 18) \div 15$

MIXED EXERCISES

EXERCISE 2L Find

1 $126 + 501 + 378$

2 $153 - 136$

3 76×9

4 $84 \div 3$

5 $350 + 8796 - 2538$

6 $8 \times 321 - 1550$

7 $35 + 86 + 94 + 27$

8 $20 \div (9 - 4) + 3$

9 Find $2 + 3 \times 4 \div 6 - 3$

10 How many packets of popcorn costing 15 p each can I buy with £1 ?

11 I buy three bars of chocolate costing 28 p each.
How much change do I get from £1 ?

12 You are asked to read pages 12 to 24 of a book for homework.
How many pages is this ?

EXERCISE 2M   Find

1 $92 + 625 + 153$ **5** $(7 + 30) \times 2 - 45$

2 $247 - 193$ **6** $382 - 792 \div 3$

3 84×8 **7** $68 - 42 + 12 \times 2$

4 $79 \div 8$ **8** $79 - 35 + 56 - 63$

9 Find $4 + 16 \times 5 \div 10 + 3$

10 How many times can 6 be taken away from 45?

11 The contents of a tin of sweets weigh 2500 grams. The sweets are divided into packets each weighing 500 grams.
How many packets of sweets can be made up?

12 Peter bought 3 lollies at 5 p each, one bar of chocolate at 25 p and a drink at 23 p.
How much change did he get from a £1 coin?

NUMBER PUZZLES

1 Copy the following sets of numbers. Put $+ - \times$ or $\div$ in each space so that the calculations are correct.

a $9 \square 4 = 5$ **e** $5 \square 4 \square 6 = 3$
b $7 \square 3 = 21$ **f** $8 \square 3 \square 4 = 1$
c $28 \square 4 = 7$ **g** $3 \square 4 \square 2 = 9$
d $8 \square 2 = 4$ **h** $2 \square 1 \square 3 = 6$

2 My calculator has an odd fault. The $+$ button multiplies and the $\times$ button adds.

a When I press $7 + 5$, what number shows on the display?

b Before I realised what the fault was, I got 21 when I tried to add two numbers.
What were the two numbers?

c I pressed $2 + 8 \times 5$. What number showed in the display?

d Find a calculation using three or four numbers for which my calculator will give the correct answer.

3 Solve this cross-number puzzle.

Across	1	$127 - 64$
	3	$44 + 73 - 58$
	4	6×53
	8	330×41
	11	$9 \times 10 - 9$
	12	The next number after 40

Down	2	$3 \times 13 - 6$
	3	$464 \div 8$
	5	$625 \div 5$
	6	74×7
	7	9×89
	9	$5 \times 8 - 27 \div 3$
	10	$2 \times 19 - 4$

INVESTIGATIONS

1 Elsie has some tennis balls and some bags to keep them in. If 9 balls are put into each bag, one ball is left over. If 11 balls are put into each bag, one bag is empty.

How many tennis balls and how many bags are there?

Explain your thinking.

2 **a** Multiply 123 456 789 by 3 and then multiply the result by 9. What do you notice?

b Repeat part **a** multiplying first by a different number less than 9.

Do this twice, without using a calculator..

c Is there a rule for predicting the answer when 123 456 789 is multiplied by one of the numbers 2, 3, 4, 5, 6, 7, or 8, and the result is multiplied by 9? If you find one, write it down and test it.

d Now try using your calculator. What do you notice?

COLLECTING AND DISPLAYING DATA

The news and advertising media bombard us with information. For example:

- Two in every five pupils do not have anything to eat before they go to school.
- Children between the ages of 8 and 12 spend an average of ten hours a week playing computer games.
- Air pollution in town centres is up to six times the recommended levels.

Statements like these are the result of a process that involves collecting information and organising it. The process starts with someone asking a question.

The first example probably started with the question 'How many children arrive at school without having had anything to eat?' It was probably answered by asking some children what they had eaten that day before coming to school and then organising the information collected.

Asking questions is easy, but deciding what information needs to be collected so that they can be answered is not always so straightforward. The third example involved collecting measurements of air pollution. But what is air pollution? Where was it measured? How often was it measured?

It is important to know precisely what information needs collecting and how much information to collect.

EXERCISE 3A   Discuss what information is needed to answer these questions.

1 How much money do the pupils in my class have each week to spend as they wish?

2 Do tubes of Smarties have more red Smarties in them than any other colour?

3 In which months are most pupils in my year born?

4 Does this dice give more sixes than any other score?

5 What pets do the other members of my class have?

6 Are left-handed people taller than right-handed people?

7 Do people with large feet also have large hands?

8 How many books do my class-mates read each week?

9 Are people who do not eat green vegetables less healthy than those who do?

10 Decide on a question you would like to ask and state what information you need to collect.

MAKING LISTS

If we want to know how many brothers and sisters each member of a class has, we could ask each pupil and write the information down in a list.

This is a list of the number of brothers and sisters of the thirty pupils in Class 7D. Each figure represents the number of brothers and sisters of one member of the class.

$$
\begin{array}{cccccccccc}
0 & 1 & 1 & 3 & 0 & 1 & 2 & 1 & 0 & 1 \\
1 & 3 & 1 & 0 & 1 & 2 & 3 & 1 & 1 & 0 \\
1 & 0 & 2 & 0 & 1 & 1 & 1 & 2 & 4 & 1
\end{array}
$$

MAKING A FREQUENCY TABLE

To make sense of the numbers that we have collected we need to put them into order. One way of doing this is to make a *frequency table*. The frequency tells us how many of each type of item there are.

There are five different items in our list: 0, 1, 2, 3 and 4.

We start by making a table like the one below. Then we work down the columns in the list, making a tally mark, /, in the tally column next to the appropriate item.
(To make it easier to count up afterwards we can group the tally marks into fives, like this: $\cancel{||||}$ //, or like this ☑∟)

Number of brothers and sisters	Tally	Frequency
0	$\cancel{||||}$	
1	$\cancel{||||}$ //	
2	//	
3	//	
4		
	Total	

Next we count up the tally marks and write the total in the frequency column.

Lastly, we total the frequency column to check that the total number of items recorded in the table is the same as the number of items in the list.

EXERCISE 3B

1 The table above is incomplete; copy it without the tally marks. Now make the tally marks afresh from the original information and complete the table.

2 A box contains bags of crisps. Some are plain salted (S), some are salt and vinegar (V), some are cheese and onion (C) and some are prawn cocktail (P). The bags are taken out of the box one at a time and the flavour of each bag is written down in a list.

$$
\begin{array}{llllllll}
P & S & S & V & C & S & P & P & S \\
S & P & V & S & C & C & P & V & S \\
P & S & V & C & C & C & V & P & V \\
V & S & S & C & S & S & P & S & S \\
\end{array}
$$

Make a frequency table for this list like the one on page 39.

3 This is a list of the favourite colours of some children, chosen from red (R), green (G), blue (B), yellow (Y) and pink (P).

$$
\begin{array}{llllllllll}
R & R & P & Y & R & Y & P & G & R & B & R \\
P & Y & Y & R & B & Y & R & R & R & Y & Y \\
R & Y & B & G & Y & R & P & P & R & R & Y \\
Y & R & R & R & P & Y & Y & B & Y & G & R \\
\end{array}
$$

Make a frequency table for this list.

4 This is a list of the shoe sizes of a group of children.

$$
\begin{array}{lllllllllll}
26 & 25 & 27 & 25 & 23 & 26 & 24 & 24 & 27 & 24 & 25 \\
24 & 23 & 25 & 24 & 24 & 23 & 26 & 23 & 24 & 25 & 25 \\
23 & 24 & 25 & 25 & 24 & 23 & 24 & 26 & 25 & 24 & 23 \\
22 & 24 & 23 & 27 & 23 & 25 & 24 & 24 & 25 & 23 & 24 \\
\end{array}
$$

Make a frequency table for this list.

OBSERVATION SHEETS

We can sometimes avoid having to sort information afterwards by collecting it straight onto a specially designed sheet that sorts the information as we collect it. This is called an observation sheet. The simplest observation sheet is a tally chart.

EXERCISE 3C

1 Copy the tally chart below and use it to record the number of heads that show when two coins are tossed twenty times.

Number of heads	Tally	Frequency
0		
1		
2		

2 This observation sheet can be used to collect information so that we can compare shoe size and hand span of several people.

Name	Shoe size	Hand span (cm)
Alice John Rajan Shona		

Explain why a tally chart would not be a suitable observation sheet for gathering this information.

3 Mrs Edwards wants to know what games equipment each member of her class has in school. She uses a sheet like the one below.

Name	Trainers	Games shirt	Tracksuit trousers	Tracksuit top
Ben	✓	✗	✓	✓

a Design a sheet on which similar information about your class could be gathered.

b Mrs Edwards just wants to know how many pupils in her class have each item in school that day. Design an appropriate sheet for just this purpose.

4 Ashley wants to record the number of people in each car passing the bus stop while he waits for the bus to school each morning. Use your personal experience to decide whether it is sensible to use a tally sheet or to write down the numbers in an unordered list and sort them later.
Give reasons for and against each option.

5 Design an observation sheet to record

a the number of letters in each word on this page

b the different types of book that are in each pupil's bag when they arrive at school tomorrow morning

c the number of pencils, rubbers, rulers, calculators, and pens brought by all the pupils to your next maths class

d the number of red cars passing a bus stop each minute.

**MAKING A BAR
CHART**

A bar chart shows frequencies very clearly.

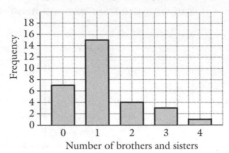

Number of brothers and sisters

This bar chart uses the frequency table in question 1 of **Exercise 3B** and shows the frequencies of the numbers of brothers and sisters of the pupils in a class.

Notice that the vertical axis is used for frequency. The different kinds of item are on the horizontal axis.

Notice also that the bars are all the same width. It does not matter what width you choose to make the bars, provided that all bars are the same width. In this diagram there is a space between the bars, but the bars can touch if we want them to.

EXERCISE 3D

1 Peter and Rachel did a survey of the types of vehicle passing the school gate one lunch hour and produced the following frequency table.

Type of vehicle	Bicycle	Motorbike	Car	Lorry
Frequency	4	10	25	16

a How many vehicles passed the school gate altogether?

b Which was the most common type of vehicle?

c Draw a bar chart to show the information.

2 This frequency table shows the results of a survey on children's opinions about the quality of school dinners.

Opinion	Very Good	Good	Satisfactory	Poor	Very poor
Frequency	2	12	20	10	8

a How many people were asked their opinion on school dinners?

b Draw a bar chart to show the information.

3 Use the frequency table that you made for question **2** of **EXERCISE 3B** for this question.

a Which is the most common type of crisp in the box?

b Draw a bar chart to show the information.

4 Use the frequency table that you made for question **3** of
EXERCISE 3B for this question.

a Which was the most commonly chosen colour?

b Draw a bar chart to show the information in the frequency table.

**USING BAR
CHARTS**

Bar charts can come in other forms. Sometimes the bars are lines (these
are called bar line graphs). Sometimes the bars are horizontal.

When you look at a bar chart to find information from it, always read the
labels on the axes carefully.

EXERCISE 3E

1 **Pets owned by members of Class 7D**

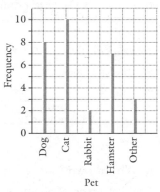

Use the bar chart to answer the following questions.

a What is the most popular pet?

b How many dogs were counted?

c What is the total number of pets owned by members of Class 7D?

2 **Marks in a maths test**

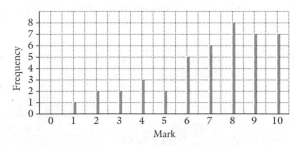

a How many pupils had a mark of 8?

b What was the lowest mark given and how many pupils got it?

c What was the most common mark?

d How many pupils had a mark of 7 or more?

3

Favourite subject from the school timetable

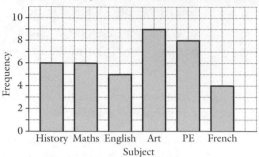

a How many children chose history as their favourite subject?

b What was the most popular subject?

c What was the least popular subject?

4

Population in five towns (to the nearest thousand)

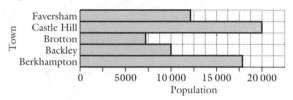

a Which town has the largest population?

b What is the population of Backley?

c Which town has the smallest population? What is the population of this town?

READING BAR CHARTS

Bar charts can be used for information other than frequencies. When you look at a bar chart, read the labels on the axes.

The bars are usually vertical but can be horizontal.

EXERCISE 3F

1 This bar chart gives rough guide-lines on the distances that should be allowed between moving cars.

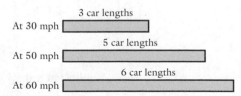

a Roughly how far should a car travelling at 70 mph keep from the car in front?

b What rule has been used to decide the safe distance?

c Write two sentences on why this rule is only a rough guide.

2   **Cost of fuel in an average home with central heating**

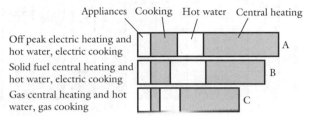

No numbers are given but we can get an idea about the relative costs.

a Which is the most expensive method overall?

b Which is the cheapest?

c Which is the most expensive method of producing hot water?

d Which is the cheapest method of cooking?

3   **Average daily hours of sunshine in Aberdeen and Margate**

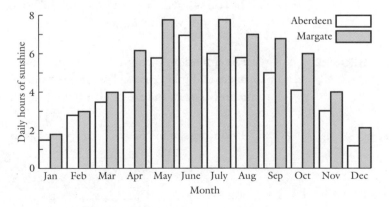

a Is there more sunshine in Margate or in Aberdeen?

b Which month is the sunniest in both towns?

c Which month has the least sunshine in each town?

4 This graph shows the number of women members of parliament elected at general elections. (From *Social Trends 22*, CSO)

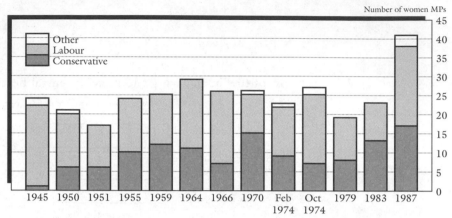

Source: House of Commons Research Division

a How many women Conservative MPs were elected in 1983?

b Were there more Labour or Conservative women MPs elected in 1987?

c In which general election were the highest number of women Labour MPs elected?

d Redraw this bar chart so that it looks similar to the bar chart for question **3**. There will be three columns for each year.

e You now have two graphs showing the same information but in different ways. Which would you find more useful if

  i you wanted to see how the total numbers of women in the House varied since 1945

  ii you wanted to compare the numbers of Conservative and Labour women MPs?

PICTOGRAPHS

To attract attention, pictographs are often used on posters and in newspapers and magazines. The best pictographs give numerical information as well; the worst give the wrong impression.

EXERCISE 3G

1 **Road deaths in the past four years at an accident black spot**

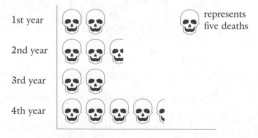

a Give an estimate of the number of deaths in each year.

b What message is the poster trying to convey?

c How effective do you think it is?

2   **The most popular subject among first year pupils**

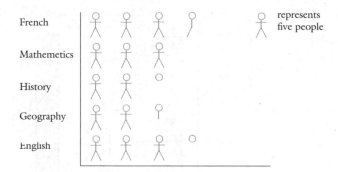

a Which is the most popular subject?

b How many pupils chose each subject and how many were asked altogether?

c Is this a good way of presenting the information? Give a reason for your answer.

3 This bar chart comes from an advertisement showing the consumption of Fizz lemonade.

a What does this show about the consumption of lemonade? What does it not show?

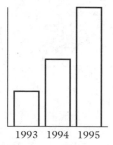

1993 1994 1995

It was decided to change from a bar chart to a pictograph for the next advertisement.

b This looks impressive but it could be misleading. Why?

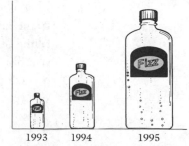

1993   1994   1995

PRACTICAL
WORK

1 This is an individual activity.

This bar chart is reproduced from *Social Trends 22*, published by The Government Statistical Service.

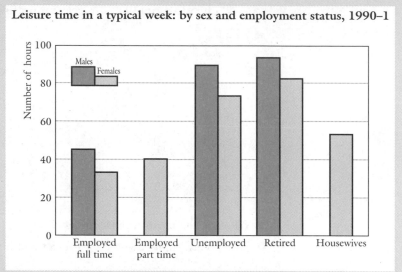

Leisure time in a typical week: by sex and employment status, 1990–1

How does your weekly leisure time compare with these figures?

Your answers should include

a a definition of what you consider to be leisure time

b a record sheet showing the number of hours spent on leisure by you for one week

c a copy of the chart above including a bar to represent your time

d an estimate of how representative your week is of your usual leisure activities and how you could improve the reliability of your estimate.

2 This is a group activity that can be done with the whole class.

This chapter starts with a statement that 2 in every 5 children do not have anything to eat before coming to school.
Is this true of your class?
Your answer should include

a the question that you ask the members of your class

b the observation sheet that you collect replies on

c a bar chart illustrating the results

d a conclusion

e a report of any difficulties encountered and suggestions as to how these could be overcome.

NUMBER AND PATTERNS

A manufacturer has three customers who want large weekly deliveries of his packs of paper plates. The first customer wants 480 packs a week, the second 960 and the third 288. The manufacturer would like to make up sealed cartons, each with the same number of packs, so that these orders can each be fulfilled by delivering a whole number of sealed cartons.

- The number of packs needed for each order, i.e. 480, 960 and 288, is an even number, so sealed cartons with two packs would be all right but it would be far more convenient to have more packs in each carton. To supply 480 packs, the number of packs in each sealed carton must divide exactly into 480. The same number must divide exactly into 960 and 288 if the other two orders can also be delivered in sealed cartons. To make the best use of the sealed cartons, this number has to be as large as possible.

The church at Bracksford has a peel of four bells. No. 1 bell rings every 5 seconds, No. 2 bell every 6 seconds, No. 3 bell every 7 seconds and No. 4 bell every 8 seconds. They are first tolled together. How can we find out how long it will be before they all sound together again?

- One way of doing this is to keep a running total of the times for each bell,

	0	5	10	15	20	25	30	35
No. 1								
No. 2								
No. 3								
No. 4								

and keep on extending these lines until the four match up.
We can see that we are looking for the smallest number that 5, 6, 7 and 8 will divide into exactly.

These and similar problems can be solved quite easily if we understand the properties of different types of numbers.

EXERCISE 4A

Discuss how you might solve the following problems.

1 Sandra wants to choose a group of players for a marching band competition. At certain times in the competition she wants them to march in equal rows of 5, and at other times in equal rows of 6. She has 65 players altogether.

What is the smallest number of players she should choose?

What is the largest number she can use?

Could she use all of them?

2 A plant breeding station wishes to divide a rectangular field measuring 558 m by 450 m into squares so that different varieties of plants can be grown under different conditions.

How can they decide on the size of the largest possible square?

To solve the problems in this exercise you have probably realised that we need to look for numbers that divide exactly into other numbers.

FACTORS

The number 2 is a *factor* of 12, since 2 will divide exactly into 12.

The number 12 may be expressed as the product of two factors in several different ways, namely: 1×12, 2×6 or 3×4.

EXERCISE 4B

Express each of the following numbers as the product of two factors in as many ways as you can.

1	18	**5**	30	**9**	48	**13**	80
2	20	**6**	36	**10**	60	**14**	96
3	24	**7**	40	**11**	64	**15**	144
4	27	**8**	45	**12**	72	**16**	160

MULTIPLES

12 is a *multiple* of 2 since 12 contains the number 2 a whole number of times. Any number that 2 divides into exactly is a multiple of 2, e.g. 2, 4, 6, 8, ... are all multiples of 2.

EXERCISE 4C

1 Write down all the multiples of 3 between 20 and 40.

2 Write down all the multiples of 5 between 19 and 49.

3 Write down all the multiples of 7 between 25 and 60.

4 Write down all the multiples of 11 between 50 and 100.

5 Write down all the multiples of 13 between 25 and 70.

PRIME NUMBERS

Some numbers can be expressed as the product of two different factors in one way only.

For example, the only factors of 3 are 1 and 3 and the only factors of 5 are 1 and 5. Any number bigger than 1 that is of this type is called a *prime number*. Note that 1 is not a prime number because it does not have two different factors.

EXERCISE 4D

1 Which of the following numbers are prime numbers?

$$2, \quad 3, \quad 4, \quad 5, \quad 6, \quad 7, \quad 8, \quad 9, \quad 10, \quad 11, \quad 12$$

2 Write down all the prime numbers between 20 and 30.

3 a Write all the numbers from 1 to 100 in a table as shown below

1	2	3	4	5	6	7	8	9	10
11	12	13	14	15	16	17	18	19	20
21	22	23	24	25	26	27	28	29	30
31	32	33	34	35	36	37	38	39	40
41	42	43	44	45	46	47	48	49	50
51	52	53	54	55	56	57	58	59	60
61	62	63	64	65	66	67	68	69	70
71	72	73	74	75	76	77	78	79	80
81	82	83	84	85	86	87	88	89	90
91	92	93	94	95	96	97	98	99	100

b Cross out 1, which is not a prime number.

c Do not cross out 2 because it is a prime number but cross out all the multiples of 2.

d Do not cross out 3 because it is a prime number but cross out all the multiples of 3.

e Do the same for 5 and repeat what you have been doing until there are no further multiples left to be crossed out.

f List the numbers that have not been crossed out. These are the prime numbers less than 100.

4 Which of the following numbers are prime numbers?

$$41, \quad 57, \quad 91, \quad 101, \quad 127$$

5 Find as many prime numbers as you can between 100 and 200.

6 Write each even number from 8 to 20 as the sum of two odd primes.

7 Write each odd number between 10 and 30 as the sum of three odd primes.

8 Are the following statements true or false? Give reasons for your answers.

a All the prime numbers are odd numbers.

b All odd numbers are prime numbers.

c All prime numbers between 10 and 100 are odd numbers.

d The only even prime number is 2.

e There are six prime numbers less than 10.

9 a List all the prime numbers less than 50.

b
```
                1
             2     5
          8     9    19
       11     5   17   12
        7    3    9   13   14
     38  13   33    7   28   36
    15  10   19    5    9   11   18
   5   47    8   23   14   11   26   13
```

Begin at the top and move down through the pyramid until you get to the bottom row. You must go through one number in each row but, apart from 1, this number must be a prime and must be next to the number you have just passed through. For example, if you are at 5 in the second row from the top the only choice you have next is 9 or 19.

Is there more than one route?

10
```
                  7
                3   5
              4   12   2
            5   8   13   46
          2   24   42   41   34
        19   43   37   45   31   21
      18   28    3   32   11   27   17
     7   11   16    5   28   47   19    5
   12   31   17    7   19   36   11   21   43
    9    5   15   11   12   13   14   19    6   19
```

Begin in the bottom row and move up, one row at a time, until you get to the top. You can pass through one number only in each row but this number must be a prime. As you go from one row to the next you must go to an adjacent number. For example, if you start at 13 in the bottom row you must go next to 19 or 36.

Is there more than one route?

INDEX NUMBERS   The accepted shorthand way of writing $2 \times 2 \times 2 \times 2$ is 2^4. We read this as '2 to the power 4' or '2 to the four'. The 4 is called the *index*. Hence $16 = 2 \times 2 \times 2 \times 2 = 2^4$ and similarly $3^3 = 3 \times 3 \times 3 = 27$.

EXERCISE 4E   Write the following products in index form.

1 $2 \times 2 \times 2$ **6** $3 \times 3 \times 3 \times 3 \times 3 \times 3$

2 $3 \times 3 \times 3 \times 3$ **7** $13 \times 13 \times 13$

3 $5 \times 5 \times 5 \times 5$ **8** 19×19

4 $7 \times 7 \times 7 \times 7 \times 7$ **9** $2 \times 2 \times 2 \times 2 \times 2 \times 2 \times 2$

5 $2 \times 2 \times 2 \times 2 \times 2$ **10** $6 \times 6 \times 6 \times 6$

Find the value of 3^5.

$$3^5 = 3 \times 3 \times 3 \times 3 \times 3$$
$$= 243$$

$3 \times 3 \times 3 \times 3 \times 3 = 9 \times 3 \times 3 \times 3$
$\qquad\qquad\qquad = 27 \times 3 \times 3$
$\qquad\qquad\qquad = 81 \times 3$
$\qquad\qquad\qquad = 243$

Find the value of

11 2^5 **13** 5^2 **15** 3^2 **17** 3^4

12 3^3 **14** 2^3 **16** 7^2 **18** 2^4

Express 125 in index form.

$$125 = 5 \times 25$$
$$= 5 \times 5 \times 5$$
$$= 5^3$$

Express the following numbers as powers of prime numbers.

19 4   **21** 8   **23** 49   **25** 32

20 9   **22** 27   **24** 25   **26** 64

PRODUCTS OF PRIME NUMBERS

We can now write any number as the product of prime numbers in index form. Consider the number 108.

$$108 = 12 \times 9$$
$$= 4 \times 3 \times 9$$
$$= 2 \times 2 \times 3 \times 3 \times 3$$

i.e. $\qquad 108 = 2^2 \times 3^3$

Therefore 108 expressed as the product of prime numbers or factors in index form is $2^2 \times 3^3$.

Similarly
$$441 = 9 \times 49$$
$$= 3 \times 3 \times 7 \times 7$$
$$= 3^2 \times 7^2$$

EXERCISE 4F

> Write the product $2 \times 3 \times 3 \times 3 \times 2$ in index form.
>
> > Remember that we can multiply a string of numbers in any order so we can arrange these to bring the 2s together.
>
> $2 \times 3 \times 3 \times 3 \times 2 = 2 \times 2 \times 3 \times 3 \times 3$
> $\qquad\qquad\qquad\quad = 2^2 \times 3^3$

Write the following products in index form.

1 $2 \times 2 \times 7 \times 7$

2 $3 \times 3 \times 3 \times 5 \times 5$

3 $5 \times 5 \times 5 \times 13 \times 13$

4 $2 \times 3 \times 3 \times 5 \times 2 \times 5$

5 $2 \times 2 \times 3 \times 2 \times 3 \times 5 \times 5$

6 $3 \times 11 \times 11 \times 2 \times 2$

7 $7 \times 7 \times 7 \times 3 \times 5 \times 7 \times 3$

8 $13 \times 5 \times 13 \times 5 \times 13$

9 $3 \times 5 \times 5 \times 3 \times 7 \times 3 \times 7$

10 $2 \times 3 \times 2 \times 5 \times 3 \times 5$

> Find the value of $2^3 \times 3^2$
>
> $2^3 \times 3^2 = 2 \times 2 \times 2 \times 3 \times 3 = 72$

Find the value of

11 $2^2 \times 3^3$

12 $3^2 \times 5^2$

13 $2^4 \times 7$

14 $2^2 \times 3^2$

15 $2^2 \times 3^2 \times 5$

16 $2 \times 3^2 \times 7$

FINDING PRIME FACTORS

The following rules may help us to decide whether a given number has certain prime numbers as factors.

A number is divisible

by 2 if the last figure is even
by 3 if the sum of the digits is divisible by 3
by 5 if the last figure is 0 or 5.

EXERCISE 4G

Is 446 divisible by 2?

Since the last figure is even, 446 is divisible by 2.

Is 1683 divisible by 3?

The sum of the digits is $1 + 6 + 8 + 3 = 18$, which is divisible by 3.
Therefore 1683 is divisible by 3.

Is 7235 divisible by 5?

Since the last digit is 5, 7235 is divisible by 5.

---

1 Is 525 divisible by 3?

2 Is 747 divisible by 5?

3 Is 2931 divisible by 3?

4 Is 740 divisible by 5?

5 Is 543 divisible by 5?

6 Is 1424 divisible by 2?

7 Is 9471 divisible by 3?

8 Is 2731 divisible by 2?

Is 8820 divisible by 15?

8820 is divisible by 5 since it ends in 0.
8820 is divisible by 3 since $8 + 8 + 2 = 18$ which is divisible by 3.

8820 is therefore divisible by both 5 and 3,
∴ it is divisible by 5×3, i.e. 15.

---

9 Is 10 752 divisible by 6?

10 Is 21 168 divisible by 6?

11 Is 30 870 divisible by 15?

12 Is 6540 divisible by 20? Explain your reasoning.

EXPRESSING A
NUMBER IN
PRIME FACTORS

We can find the prime factors of small numbers like 12 and 18 easily, but bigger numbers need to be dealt with in a more orderly way.
To express a number in prime factors start by trying to divide by 2 and keep on until you can no longer divide exactly by 2. Next try 3 in the same way, then 5 and so on for each prime number until you are left with 1.

EXERCISE 4H

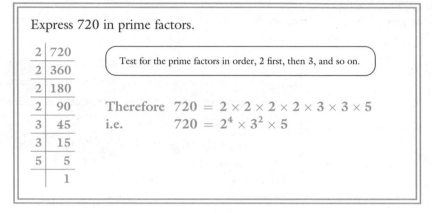

Express 720 in prime factors.

2	720
2	360
2	180
2	90
3	45
3	15
5	5
	1

Test for the prime factors in order, 2 first, then 3, and so on.

Therefore $720 = 2 \times 2 \times 2 \times 2 \times 3 \times 3 \times 5$
i.e. $720 = 2^4 \times 3^2 \times 5$

Express each of the following numbers in prime factors.

1 24 **3** 63 **5** 136 **7** 216 **9** 405

2 28 **4** 72 **6** 84 **8** 528 **10** 784

COMMON
FACTORS

The manufacturer referred to at the beginning of this chapter can solve his problem if he can find the largest number that will divide exactly into 480, 960 and 288.

Two or more numbers may have the same factor. This is called a common factor.
For example, 7 is a factor of 14, 28 and 42. Sometimes it is useful to find the highest factor that is common to a set of numbers. In this case the highest factor that will divide exactly into 14, 28 and 42 is 14.

EXERCISE 4I

Find the largest whole number that will divide exactly into all the given numbers.

1 9, 12 **5** 22, 33, 44 **9** 14, 42

2 8, 16 **6** 21, 42, 84 **10** 39, 13, 26

3 12, 24 **7** 25, 35, 50, 60 **11** 15, 30, 45, 60

4 25, 50, 75 **8** 36, 44, 52, 56 **12** 10, 18, 20, 36

13 Solve the manufacturer's problem stated at the beginning of the chapter.

COMMON MULTIPLES

The bell ringing problem discussed at the beginning of the chapter can be solved if we find the smallest number that 5, 6, 7 and 8 will divide into exactly.

The smallest common multiple of two or more numbers is the smallest number that the numbers divide into exactly. The smallest number that 5 and 12 will divide into is 60 and the smallest number that 3, 6 and 5 will divide into is 30. (Notice that 3 divides exactly into 6 so we need not worry about the 3.)

EXERCISE 4J

Find the smallest number that can be divided exactly by 15 and by 20.

When the required number is not obvious, we try multiples of 15 until we find one that 20 divides into exactly; i.e. 15, 30, 45, 60.

60 is the smallest number that 15 and 20 divide into exactly.

Find the smallest number that all the given numbers will divide into exactly.

1 3, 5	**5** 12, 16, 24	**9** 9, 12
2 6, 8	**6** 4, 5, 6	**10** 10, 15, 20
3 5, 15	**7** 9, 12, 18	**11** 9, 12, 36
4 3, 9, 12	**8** 18, 27, 36	**12** 6, 7, 8

13 Now solve the bell ringing problem.

EXERCISE 4K

Some problems in this exercise can be solved by finding factors, others by finding multiples.

1 What is the smallest sum of money that can be made up of an exact number of 20 p pieces or of 50 p pieces?

2 Find the least sum of money into which 24 p, 30 p and 54 p will divide exactly.

3 Find the smallest length that can be divided exactly into equal sections of length 5 m or 8 m or 12 m.

4 The light at one lighthouse flashes every 8 seconds while the light at another lighthouse flashes every 15 seconds. They flash together at midnight.

What is the time when they next flash together?

5 If I go up a flight of stairs two at a time I get to the top without any being left over. If I then try three at a time and again five at a time, I still get to the top without any being left over.

Find the shortest flight of stairs for which this is possible.

How many would remain if I were to go up this flight seven at a time?

6 A room measures 450 cm by 350 cm.

Find the side of the largest square tile that can be used to tile the floor without any cutting.

7 Two cars travel around a Scalextric track, one completing the circuit in 6 seconds and the other in $6\frac{1}{2}$ seconds.

If they leave the starting line together how long will it be before they are again side by side?

8 Find the largest number of children who can share equally 126 oranges and 147 bananas.

9 A gear wheel with 30 teeth drives another wheel with 65 teeth. A certain pair of teeth are touching when the wheels start.

How many times must each wheel turn before the same two teeth touch each other again?

10 In the first year of a large comprehensive school it is possible to divide the pupils into equal-sized classes of either 24 or 30 or 32 and have no pupils left over.

Find the size of the smallest entry that makes this possible.

How many classes will there be if each class is to have 24 pupils?

NUMBER
PATTERNS

Two very simple number patterns are: 1, 2, 3, 4, 5, . . . and 2, 4, 6, 8, . . . that is, the pattern of odd numbers and the pattern of even numbers.

These patterns are continued by adding a fixed amount (2) to each term to get the next term.

EXERCISE 4L

In questions **1–6** each pattern is formed, either by adding a fixed amount to, or by subtracting a fixed amount from, each term to get the next one. Write down this difference stating clearly whether the quantity is added or subtracted.

1 2, 6, 10, 14   **4** 6, 17, 28, 39

2 5, 10, 15, 20   **5** 20, 17, 14, 11

3 4, 11, 18, 25   **6** 62, 57, 52, 47

In questions **7–12** find the difference between consecutive terms in the pattern and hence write down the next three terms of each pattern.

7 1, 5, 9, 13   **10** 43, 36, 29, 22

8 10, 20, 30, 40   **11** 5, 13, 21, 29

9 60, 55, 50, 45   **12** 6, 18, 30, 42

You want Shelley to write down the sequence of numbers 4, 11, 18, 25, 32 . . . by giving her the first term followed by an instruction. What would you tell Shelley to do?

I would tell Shelley
'Write down the sequence of numbers you get by starting with 4 and then adding 7 each time'.

You note that each term is found by adding 7 to the previous term.

For questions **13–18** write down a general instruction that would give the pattern.

13 10, 25, 40, 55   **16** 100, 91, 82, 73

14 7, 10, 13, 16   **17** 63, 55, 47, 39

15 8, 15, 22, 29   **18** 8, 21, 34, 47

RECTANGULAR NUMBERS

Any number that can be shown as a rectangular pattern of dots is called a rectangular number.

24 is a rectangular number e.g.:

5 is not a rectangular number since a line of dots is not a rectangle, i.e. • • • • •

EXERCISE 4M

In questions **1–5** draw dot patterns for any of the following numbers that are rectangular numbers.

1 8 **2** 15 **3** 23 **4** 27 **5** 35

6 Show in two different ways that 12 is a rectangular number.

7 Which of the numbers between 2 and 20 are not rectangular numbers? What special name do we give to these numbers?

SQUARE NUMBERS

If the number of dots representing a number can be arranged as a square the number is called a square number.

For example, 9 and 25 are square numbers.

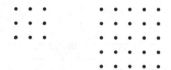

Note that 1 is also a square number.

EXERCISE 4N

1 **a** Write down the first six square numbers.

b What is **i** the tenth square number
ii the twelfth square number?

2 Look at the pattern.

$$
\begin{array}{c}
1 \\
1\ 2\ 1 \\
1\ 2\ 3\ 2\ 1 \\
1\ 2\ 3\ 4\ 3\ 2\ 1
\end{array}
$$

What total do you get for each line in this pattern?
Are all these totals rectangular numbers and/or square numbers?

3 Repeat question **2** for the pattern formed by adding the odd numbers.

$$
\begin{array}{l}
1 \\
1 + 3 \\
1 + 3 + 5 \\
1 + 3 + 5 + 7
\end{array}
$$

TRIANGULAR NUMBERS

If the dots are arranged in rows so that each row is one dot longer than the row above, we have a pattern of triangular numbers.

1: •   3: •• •   6: •• •• •   10: •• •• •• •

EXERCISE 4P

1 Draw a dot pattern for the next three triangular numbers after 10.

2 Without drawing a dot pattern write down the next three triangular numbers after 28.

3 What pattern do you get if you sum the numbers in each line of this pattern?

$$1$$
$$1 + 2$$
$$1 + 2 + 3$$
$$1 + 2 + 3 + 4$$
$$1 + 2 + 3 + 4 + 5$$

4 Using the text and your answers for questions **1** and **2**, write down the first 10 triangular numbers in order. Now follow these instructions.

Step 1 Choose any triangular number other than 1.

Step 2 Multiply your chosen number by the next but one triangular number. Note your answer.

Step 3 Write down the triangular number that was between the two numbers you chose.

Step 4 Multiply this number by the number that is 1 smaller.

Step 5 Compare your Step 2 and Step 4 answers.

Step 6 Repeat the previous steps for a different number.

OTHER PATTERNS

Lots of other patterns are possible apart from those we have already considered.

EXERCISE 4Q

In questions **1** to **10** write down the next three terms for each pattern.

1 1, 2, 4, 8, 16,   **4** 1, 2, 4, 7, 11,

2 5, 25, 125, 625,   **5** 5, 7, 11, 17, 25,

3 2, 6, 18, 54,   **6** 2, 3, 5, 7, 11, 13,

In some patterns the next number is found by adding the previous pair. For example 2, 4, 6, 10, 16, ...

7 4, 2, 6, 8,   **9** 2, 1, 3, 4, 7,

8 1, 2, 3, 5, 8,   **10** 1, 1, 2, 3, 5,

A sequence of patterns can also be formed from shapes.

Give the next two shapes for the following patterns.

11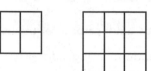

12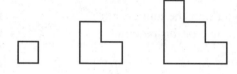

13

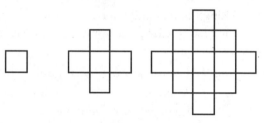

14

MIXED EXERCISE

EXERCISE 4R

1 Write $3 \times 3 \times 3 \times 3 \times 3$ in index form.

2 Find the value of **a** 4^3 **b** $3^2 \times 4^3$

3 Express 216 as the product of prime numbers in index form.

4 Is 1273 divisible by 3?

5 Find the largest whole number that will divide exactly into 21, 42 and 63.

6 Find the smallest whole number that 6, 8 and 12 will divide into exactly.

7 Find the largest number of children that can share equally 72 sweets and 54 chocolates.

8 Write down the next three terms in the pattern

a 20, 26, 32, 38, ...   **b** 1, 3, 7, 13, 21, ...

INVESTIGATIONS

1

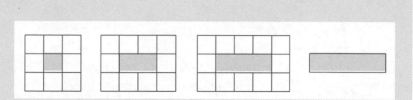

The diagrams show a pattern that starts with 1 green square surrounded by white squares. As the row of green squares increases so does the number of white squares needed to enclose it.

a Copy and complete the fourth diagram and draw the fifth and sixth diagrams.
Copy and complete the following table.

Position of diagram	Number of green squares	Number of white squares needed to surround the green squares
1	1	8
2	2	
3	3	
4	4	
5	5	
6		

b Without drawing diagrams can you say how many white squares are needed if the number of green squares is
i 10   **ii** 20 ?

c Write, in words, the connection between the number of white squares and the number of green squares in each diagram.

d Is there a connection between the number of green squares and the numbers of white squares if you arrange the green squares as nearly as possible in the shape of a square instead of a row ?
For example

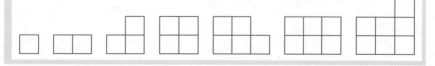

2 a Write down the next three steps for the pattern

$$4 \times 9 = 36$$
$$44 \times 9 = 396$$
$$444 \times 9 = 3996$$

b Does your method work when other similar patterns of a different digit are multiplied by 9?
Try, for example $7 \times 9,\ 77 \times 9,\ 777 \times 9, \dots$
and then $8 \times 9,\ 88 \times 9,\ 888 \times 9, \dots$
Without multiplying, write down the pattern you get when 5, 55, 555, 5555, and so on are multiplied by 9.
Check your values by doing the multiplication.
Does this pattern always follow?
Can you justify your answer?

3 a If the digits of a number are added together we have a reduced number. For example 26 reduces to 8, 167 reduces to 14 which further reduces to 5.
Investigate the reduced numbers for the numbers in the pattern

$$1, 1, 2, 3, 5, 8, 13, 21, 34, \dots$$

Add more numbers to the original pattern. If you go far enough the pattern of reduced numbers starts to repeat itself.
Why does this happen?
Can you be certain that it will continue to happen as you add more numbers to the pattern?

b Investigate the pattern of reduced numbers for the numbers

$$1, 3, 4, 7, 11, 18, 29, \dots$$

What happens when you write down alternate numbers from the pattern of reduced numbers? Do not give up too soon!
What happens if you start the pattern with the numbers 3, 1 instead of 1, 3?

c Try making up some similar patterns of your own. Find the reduced numbers for your pattern .
Do they have a pattern?

SUMMARY 1

MIXED ADDITION AND SUBTRACTION

The sign in front of a number refers to that number only.
The order in which you add or subtract does not matter, so
$1 - 5 + 8$ can be calculated in the order $1 + 8 - 5$, i.e. $9 - 5 = 4$

ROUNDING NUMBERS

To round a number to the nearest 10, look at the units; if there are 5 or more units, add one to the number of tens, otherwise leave it untouched. To round a number to the nearest hundred, look at the tens; add one to the hundreds if there are 5 or more tens, otherwise leave it untouched,

e.g. $\quad 137 = 140$ to the nearest 10
and $\quad 137 = 100$ to the nearest 100

MIXED OPERATIONS OF $\times, \div, +, -$

When a calculation involves a mixture of operations, start by calculating anything inside brackets, then follow the rule 'do the multiplication and division first',

e.g. $\quad 2 + 3 \times 2 = 2 + 6 \qquad$ and $\qquad (2 + 3) \times 2 = 5 \times 2$
$\qquad\qquad\qquad\quad = 8 \qquad\qquad\qquad\qquad\qquad\qquad = 10$

FACTORS AND MULTIPLES

A factor of a number will divide into the number exactly.
When two or more numbers have the same factor, it is called a common factor.
A multiple of a number has that number as a factor,

e.g. $\quad$ 3 is a factor of 12 and 12 is a multiple of 3

A common multiple of two or more numbers can be divided exactly by each of those numbers.

TYPES OF NUMBER

Even numbers divide exactly by 2, e.g. $6, 10, \ldots$
Odd numbers do not divide exactly by 2, e.g. $3, 7, \ldots$
A prime number has only 1 and itself as factors, e.g. 7.
Remember that 1 is *not* a prime number.

Square numbers can be drawn
as a square grid of dots, e.g. 9:

Rectangular numbers can be
drawn as a rectangular grid of dots, e.g. 6:

Triangular numbers can be drawn
as a triangular grid of dots, e.g. 6:

INDEX NUMBERS   The small 2 in 3^2 is called an index and it tells you how many 3s are multiplied together,

e.g. 3^2 means 3×3 and 3^5 means $3 \times 3 \times 3 \times 3 \times 3$

REVISION
EXERCISE 1.1
(Chapters 1 and 2)

1 Write the number

   a three hundred and seventy-two in figures   **b** 4076 in words

2 Find, without using a calculator

   a $56 + 31$ **b** $199 - 48$ **c** $114 + 58 + 47$

3 Find, without using a calculator   **a** $496 + 265$ **b** $608 - 399$

4 Find the distance round the edge of this template.

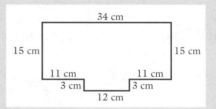

5 Helen was due to have a history lesson that lasted 55 minutes. She was 7 minutes late in arriving and had to leave 15 minutes early to go for a dental appointment.
How long was she in the lesson?

6 Find, without using a calculator

   a 574×3 **b** 76×25 **c** 1300×360

7 a Find, giving the remainder if there is one

    i $427 \div 5$ **ii** $1035 \div 23$ **iii** $1624 \div 36$

   b Find $10 + 4 \times 3 - 12 + 2$

8 In a concert hall there are 24 rows in Tier 2 with 18 seats in each row.
How many people can be seated in Tier 2?

9 Estimate the value of 197×38 and then use a calculator to find the exact value.

10 How many 13-seater minibuses are needed to transport 50 children to a match?
How many spare seats are there?

**REVISION
EXERCISE 1.2
(Chapters 3 and 4)**

1 Gemma counted the numbers of people in the cars that passed her home one lunch time. The numbers she recorded were

```
1  1  1  1  4  1  1  1
2  4  2  2  1  3  2  1
1  1  3  1  1  1  1  2
1  2  1  1  2  1  1  3
```

a How many cars did she count?

b How many people did she count altogether?

c Make a frequency table for this list.

2 The bar chart shows the number of letters delivered by a postman to each house in a street one morning.

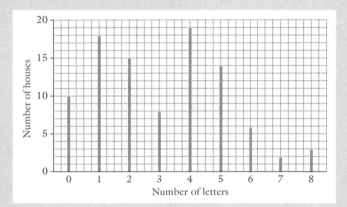

a What was the most common number of letters delivered?

b How many houses received 4 letters?

c How many houses received 4 letters or more?

d How many houses were there in the street?

e How many houses did not have a letter delivered?

f How many letters were delivered altogether?

3 Write down all the multiples of 7 between 30 and 50.

4 Write in index form   **a** $5 \times 5 \times 5$ **b** $3 \times 3 \times 3 \times 3 \times 3 \times 3$

5 Find the largest whole number that will divide exactly into

   a 14 and 21   **b** 52, 13 and 39   **c** 15, 40 and 65

6 Express as prime numbers in index form   **a** 81   **b** 64   **c** 48

7 Write down the next three terms in the pattern

   a 10, 25, 40, 55   **b** 4, 12, 20, 28

8 Explain how you can find out whether 2493 is divisible by 3 without doing the division.

9 A gear wheel with 35 teeth drives another wheel with 55 teeth. When the wheels begin to turn two particular teeth are touching. How many turns must each wheel make before the same two teeth touch each other again?

10   6, 7, 8, 12, 15, 16, 17, 19, 20, 21, 23, 25

Which of the numbers in this list are

a prime numbers   **c** triangular numbers

b square numbers   **d** rectangular numbers?

REVISION EXERCISE 1.3 (Chapters 1 to 4)

1 Find the sum of one thousand and forty-two, sixty-nine and five hundred and thirty-six.

2 A machine produced 7294 components in one hour. If 149 were found to be faulty, how many were satisfactory?

3 Find, without using a calculator
   a $460 \div 20$ **b** 504×60 **c** $12 \div 3 - 6 \div 2 + 1$

4 A can of tomatoes weighs 458 grams. What is the weight of 56 cans?

5 By writing each number correct to the nearest 10 find an approximate value for
   a $163 + 191$ **b** $732 - 65$ **c** $53 + 78 + 82$

6 What is the cost of 4 cream cakes at 47 p each?

7 Express in prime factors   **a** 56   **b** 540   **c** 3087

8 Write down the next three terms in the pattern
   a $2, 4, 6, 8, \ldots$ **b** $2, 4, 8, 16, \ldots$ **c** $2, 4, 6, 10, \ldots$

9 A sequence of numbers is formed by starting with 5 and then multiplying the last number by 6. Write down the first four numbers in the sequence.

10 The customers at a video store were placed in one of four categories man(M), woman(W), boy(B), girl(G). The list shows the information as it was collected.

G	B	G	M	G	G	B	G	B
B	W	W	G	G	G	W	M	G
M	W	B	B	M	W	B	M	W
G	W	G	B	W	W	W	B	W

a Make a frequency table for this information.

b Draw a bar chart to illustrate the information.

c How many of the customers were boys and how many girls?

d How many more women customers than men customers were there?

e How many customers were there altogether?

REVISION EXERCISE 1.4 (Chapters 1 to 4)

1 Find, without using a calculator

   a $314 - 169 + 249 - 73$ **c** $12 + 4 \times 3 \div 2 + 24 \div 3$

   b $64 - 184 - 47 + 206$

2 Ben needs to score 301 to win a game of darts. He scores 83 on his first turn and 39 on his second turn.
How many does he still have to score to win?

3 Fill in the blanks (marked with $\square$) in the following calculations.

   a $35 \times 49 = 171\square$ **c** $63 \times \square 2 = 4536$

   b $4\square \times 57 = 2394$ **d** $54 \times 78 = 4\square 12$

4 Use a calculator to find

   a 83×37 **c** 517×429

   b $416 \div 32$ **d** $435 - (56 - 37) \times 14$

5 Susan did a survey of the main hobby of each pupil in her class, including herself, and produced the following frequency table.

Type of hobby	Collecting stamps or coins	Playing a sport	Reading	Music	Computers
Frequency	10	9	2	3	4

   a How many pupils are there in Susan's class?

   b Which hobby is the most popular?

   c Draw a bar chart to show this information.

6 Write each even number between 30 and 40 as the sum of two primes.

7 Find the smallest whole number that 8, 10 and 16 will divide into exactly.

8 A No 9 bus is due every 8 minutes and a No 64 is due at the same bus stop every 10 minutes. Both buses are due at 8.30 a.m.
At what time are a No 9 bus and a No 64 bus next due to arrive at this bus stop together?

PARTS OF A WHOLE

Almost every day we refer to parts of a whole: $\frac{2}{3}$ of the pupils in my class are girls; 10% of the workforce are off sick; the length of the table is 0.9 m long.

In this chapter you are introduced to the different ways of describing parts of a whole, and the relationships between them.

THE MEANING OF FRACTIONS

Think of cutting a cake right through the middle into two equal pieces. Each piece is one half of the cake. One half is a fraction, written as $\frac{1}{2}$.

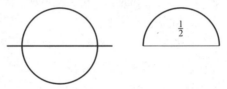

If we cut the cake into four equal pieces, each piece is one quarter, written $\frac{1}{4}$, of the cake. When one piece is taken away there are three pieces left, so the fraction that is left is three quarters, or $\frac{3}{4}$.

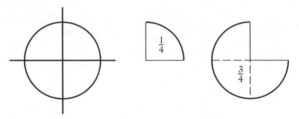

When the cake is divided into five equal slices, one slice is $\frac{1}{5}$, two slices is $\frac{2}{5}$, three slices is $\frac{3}{5}$ and four slices is $\frac{4}{5}$ of the cake.

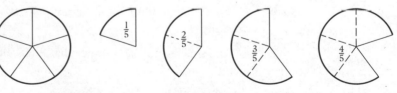

Notice that the top number in each fraction (called the *numerator*) tells you *how many* slices and the bottom number (called the *denominator*) tells you about the size of the slices.

EXERCISE 5A   In each of the following sketches, write down the fraction that is shaded.

1

4

2

5

3

6

It is not only cakes that can be divided into fractions. Anything at all that can be split up can be divided into fractions.

Write down the fraction that is shaded in each of the following diagrams.

7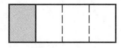

12

8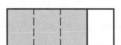

13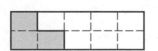

9

14

10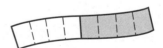

15

11

16

MIXED NUMBERS

So far we have considered *proper fractions* i.e. fractions that are less than 1, but some fractions are bigger than 1.

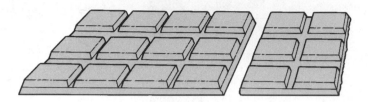

Suppose we have one and a half bars of chocolate. We write this as $1\frac{1}{2}$ bars, and $1\frac{1}{2}$ is called a *mixed number*.

Other examples of mixed numbers are $1\frac{7}{10}$, $12\frac{3}{4}$ and $24\frac{2}{5}$.

EQUIVALENT FRACTIONS

In **A**, a cake is cut into four equal pieces. One slice is $\frac{1}{4}$ of the cake.

In **B** the cake is cut into eight equal pieces. Two slices is $\frac{2}{8}$ of the cake.

In **C** the cake is cut into sixteen equal slices. Four slices is $\frac{4}{16}$ of the cake.

A   **B**   **C**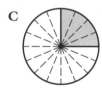

But the same amount of cake has been taken each time.

Therefore $$\frac{1}{4} = \frac{2}{8} = \frac{4}{16}$$

and we say that $\frac{1}{4}$, $\frac{2}{8}$ and $\frac{4}{16}$ are *equivalent fractions*.

Now $$\frac{1}{4} = \frac{1 \times 2}{4 \times 2} = \frac{2}{8} \qquad \text{and} \qquad \frac{1}{4} = \frac{1 \times 4}{4 \times 4} = \frac{4}{16}$$

So all we have to do to find equivalent fractions is to multiply the numerator and the denominator by the same number. For instance

$$\frac{1}{4} = \frac{1 \times 3}{4 \times 3} = \frac{3}{12}$$

and $$\frac{1}{4} = \frac{1 \times 5}{4 \times 5} = \frac{5}{20}$$

Any fraction can be treated in this way.

EXERCISE 5B

In questions **1** to **6** draw cake diagrams to show that

1 $\dfrac{1}{3} = \dfrac{2}{6}$ **3** $\dfrac{1}{5} = \dfrac{2}{10}$ **5** $\dfrac{2}{3} = \dfrac{6}{9}$

2 $\dfrac{1}{2} = \dfrac{3}{6}$ **4** $\dfrac{3}{4} = \dfrac{9}{12}$ **6** $\dfrac{2}{3} = \dfrac{8}{12}$

Fill in the missing numbers to make equivalent fractions.

a $\dfrac{1}{5} = \dfrac{3}{}$ **b** $\dfrac{1}{5} = \dfrac{}{20}$

a $\dfrac{1}{5} = \dfrac{1 \times 3}{5 \times 3} = \dfrac{3}{15}$ If $\frac{1}{5} = \frac{3}{}$ the numerator has been multiplied by 3.

b $\dfrac{1}{5} = \dfrac{1 \times 4}{5 \times 4} = \dfrac{4}{20}$ If $\frac{1}{5} = \frac{}{20}$ the denominator has been multiplied by 4.

In questions **7** to **30** copy the fractions and fill in the missing numbers to make equivalent fractions.

7 $\dfrac{1}{3} = \dfrac{2}{}$ **11** $\dfrac{2}{9} = \dfrac{4}{}$ **15** $\dfrac{1}{10} = \dfrac{10}{}$

8 $\dfrac{2}{5} = \dfrac{}{10}$ **12** $\dfrac{3}{8} = \dfrac{}{80}$ **16** $\dfrac{1}{6} = \dfrac{3}{}$

9 $\dfrac{3}{7} = \dfrac{9}{}$ **13** $\dfrac{5}{11} = \dfrac{}{22}$ **17** $\dfrac{1}{3} = \dfrac{}{12}$

10 $\dfrac{9}{10} = \dfrac{}{40}$ **14** $\dfrac{4}{5} = \dfrac{8}{}$ **18** $\dfrac{2}{5} = \dfrac{6}{}$

19 $\dfrac{3}{7} = \dfrac{}{28}$ **23** $\dfrac{4}{5} = \dfrac{}{50}$ **27** $\dfrac{4}{5} = \dfrac{}{20}$

20 $\dfrac{2}{9} = \dfrac{}{36}$ **24** $\dfrac{1}{10} = \dfrac{100}{}$ **28** $\dfrac{2}{3} = \dfrac{12}{}$

21 $\dfrac{3}{8} = \dfrac{}{800}$ **25** $\dfrac{9}{10} = \dfrac{90}{}$ **29** $\dfrac{2}{9} = \dfrac{20}{}$

22 $\dfrac{5}{11} = \dfrac{50}{}$ **26** $\dfrac{1}{6} = \dfrac{}{36}$ **30** $\dfrac{3}{8} = \dfrac{3000}{}$

Write $\frac{2}{3}$ as an equivalent fraction with denominator 24.

$$\frac{2}{3} = \frac{2 \times 8}{3 \times 8} = \frac{16}{24}$$

To change 3 to 24, we need to multiply 3 by 8, but if we multiply the denominator by 8 we must also multiply the numerator by 8.

31 Write each of the following fractions as an equivalent fraction with denominator 24.

a $\frac{1}{2}$ **b** $\frac{1}{3}$ **c** $\frac{1}{6}$ **d** $\frac{3}{4}$ **e** $\frac{5}{12}$ **f** $\frac{3}{8}$

32 Write each of the following fractions in equivalent form with denominator 45.

a $\frac{2}{15}$ **b** $\frac{4}{9}$ **c** $\frac{3}{5}$ **d** $\frac{1}{3}$ **e** $\frac{14}{15}$ **f** $\frac{1}{5}$

33 Find an equivalent fraction with denominator 36 for each of the following fractions.

a $\frac{3}{4}$ **b** $\frac{5}{9}$ **c** $\frac{1}{6}$ **d** $\frac{5}{18}$ **e** $\frac{7}{12}$ **f** $\frac{2}{3}$

34 Change each of the following fractions into an equivalent fraction with numerator 12.

a $\frac{1}{6}$ **b** $\frac{3}{4}$ **c** $\frac{6}{7}$ **d** $\frac{4}{5}$ **e** $\frac{2}{3}$ **f** $\frac{1}{2}$

35 Some of the following equivalent fractions are correct but two of them are wrong. Find the wrong ones and explain why they are wrong.

a $\frac{2}{5} = \frac{6}{15}$ **b** $\frac{2}{3} = \frac{4}{9}$ **c** $\frac{3}{7} = \frac{6}{14}$

d $\frac{4}{9} = \frac{12}{27}$ **e** $\frac{7}{10} = \frac{77}{100}$ **f** $\frac{9}{13} = \frac{18}{26}$

**SIMPLIFYING
FRACTIONS**

Think of the way you find equivalent fractions.

For example, $\dfrac{2}{5} = \dfrac{2 \times 7}{5 \times 7} = \dfrac{14}{35}$

Looking at this the other way round we see that

$$\dfrac{14}{35} = \dfrac{\cancel{7} \times 2}{\cancel{7} \times 5} = \dfrac{2}{5}$$

In the middle step, 7 is a factor of both the numerator and the denominator and it is called a common factor. To get the final value of $\frac{2}{5}$ we have 'crossed out' the common factor and this is called *cancelling*. What we have really done is to divide the top and the bottom by 7 and this *simplifies* the fraction.

When all the simplifying is finished we say that the fraction is in its *lowest terms*.

Any fraction whose numerator and denominator have a common factor (perhaps more than one) can be simplified in this way. Suppose, for example, that we want to simplify $\frac{24}{27}$. As 3 is a factor of 24 and of 27, we say

$$\dfrac{24}{27} = \dfrac{3 \times 8}{3 \times 9} = \dfrac{8}{9}$$

A quicker way to write this down is to divide the numerator and the denominator mentally by the common factor, crossing them out and writing the new numbers beside them (it is a good idea to write the new numbers smaller so that you can see that you have simplified the fraction), i.e.

$$\dfrac{\cancel{24}^{\,8}}{\cancel{27}_{\,9}} = \dfrac{8}{9}$$

EXERCISE 5C

Simplify $\dfrac{66}{176}$

$\dfrac{\cancel{66}^{\,\cancel{33}^{\,3}}}{\cancel{176}_{\,\cancel{88}_{\,8}}} = \dfrac{3}{8}$ We divided top and bottom by 2 and then by 11.

Simplify the following fractions.

1 $\dfrac{2}{6}$ **3** $\dfrac{3}{9}$ **5** $\dfrac{9}{27}$ **7** $\dfrac{5}{15}$ **9** $\dfrac{10}{20}$

2 $\dfrac{30}{50}$ **4** $\dfrac{6}{12}$ **6** $\dfrac{4}{8}$ **8** $\dfrac{12}{18}$ **10** $\dfrac{8}{32}$

11 $\dfrac{8}{28}$ **13** $\dfrac{14}{70}$ **15** $\dfrac{16}{56}$ **17** $\dfrac{36}{72}$ **19** $\dfrac{60}{100}$

12 $\dfrac{27}{90}$ **14** $\dfrac{24}{60}$ **16** $\dfrac{10}{30}$ **18** $\dfrac{15}{75}$ **20** $\dfrac{36}{90}$

INTRODUCING DECIMALS

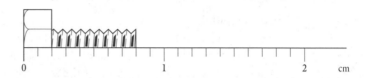

This bolt is being measured using a ruler marked in centimetres and tenths of a centimetre. The length of the bolt is $\frac{8}{10}$ of a centimetre. Other instruments have scales that are divided into tenths or hundredths as you will see if you look at a set of kitchen scales or a thermometer. Many quantities are also divided into tenths or hundredths.

- £1 is divided into 100 pence
- 1 metre is divided into 100 centimetres

Instead of writing $\frac{1}{10}$ths, $\frac{1}{100}$ths and $\frac{1}{1000}$ths we use decimal notation which is much simpler.

THE MEANING OF DECIMALS

Consider the number 426. The position of the figures indicates what each figure represents. We can write:

hundred	tens	units
4	2	6

Each quantity in the heading is $\frac{1}{10}$ of the quantity to its left: ten is $\frac{1}{10}$ of a hundred, a unit is $\frac{1}{10}$ of ten. Moving further to the right we can have further headings: tenths of a unit, hundredths of a unit and so on (a hundredth of a unit is $\frac{1}{10}$ of a tenth of a unit). For example:

tens	units		tenths	hundredths
1	6	**.**	0	2

To mark where the units come we put a point after the units position. 16.02 is 1 ten, 6 units and 2 hundredths, i.e. $16\frac{2}{100}$.

0.004 is 4 thousandths or $\frac{4}{1000}$. In this case, 0 is written before the point to help make it clear that there are no units and so that the point doesn't get lost.

units		tenths	hundredths	thousandths
0	**.**	0	0	4

EXERCISE 5D

Write **a** 34.62   **b** 0.207, in headed columns.

a   tens   units   tenths   hundredths
  34.62 =   3   4   .   6   2

b   units   tenths   hundredths   thousandths
  0.207 =   0   .   2   0   7

Write the following numbers in headed columns.

1 2.6   **5** 101.3   **9** 6.34

2 32.1   **6** 0.007   **10** 0.604

3 6.03   **7** 1.046   **11** 15.045

4 0.09   **8** 12.001   **12** 0.092

Express as a decimal   **a** $\dfrac{3}{10}$ **b** $\dfrac{13}{100}$

a $\dfrac{3}{10} = 0.3$

> The first place after the decimal point is 'tenths' so $\frac{3}{10}$ is 0.3.

b $\dfrac{13}{100} = 0.13$

> $\frac{13}{100} = \frac{10}{100} + \frac{3}{100}$, and $\frac{10}{100} = \frac{1}{10}$, so 1 goes in the first decimal place and 3 goes in the second decimal place.

Express as decimals

13 $\dfrac{1}{10}$ **17** $\dfrac{15}{100}$ **21** $\dfrac{2}{10}$

14 $\dfrac{3}{100}$ **18** $\dfrac{81}{100}$ **22** $\dfrac{5}{100}$

15 $\dfrac{7}{10}$ **19** $\dfrac{37}{100}$ **23** $\dfrac{17}{100}$

16 $\dfrac{9}{100}$ **20** $\dfrac{4}{10}$ **24** $\dfrac{38}{100}$

25 Express $\frac{2}{5}$ as a decimal. Explain your reasoning.

THE IDEA OF A PERCENTAGE

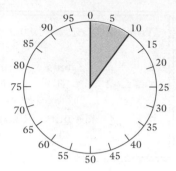

The shaded part of the cake is $\frac{10}{100}$ of the cake, or **0.1** of the cake.
Another way to describe this part of a whole is to say that 10% of the cake is shaded.

10% means '10 out of 100' and we say '10 per cent'.

(Cent comes from the Latin word for 100.)

30% means '30 out of 100' or $\frac{30}{100}$.

EXERCISE 5E

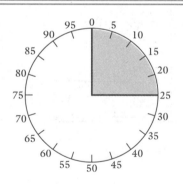

Express the shaded part of the cake as

a a percentage   **b** a decimal   **c** a fraction

a The shaded part of the cake is 25 parts out of 100 i.e. **25%**

b As a decimal $\frac{25}{100} = 0.25$

c As a fraction $\frac{25}{100}$ simplifies to $\frac{1}{4}$.

In questions **1** to **12** write down the part of the circle that is shaded as **a** a percentage **b** a decimal **c** a fraction.

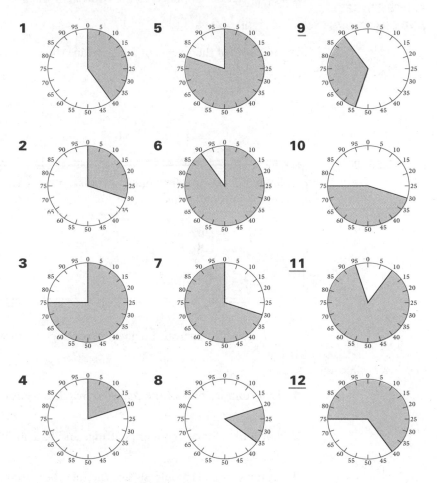

In questions **13** to **15** estimate the part of the circle that is shaded. Give your answer as **a** a percentage **b** a decimal **c** a fraction.

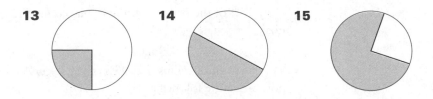

For questions **16** to **20** draw a circle and calibrate it like the circle in the text. Shade the part of the circle asked for.

16 $\frac{3}{10}$ **17** 0.6   **18** 60%   **19** 0.4   **20** $\frac{7}{10}$

CHANGING BETWEEN PERCENTAGES AND FRACTIONS

We know that 20% of the cars in a car-park means $\frac{20}{100}$ of the cars there. Now $\frac{20}{100}$ can be simplified to the equivalent fraction $\frac{1}{5}$,

i.e. $20\% = \frac{1}{5}$.

Similarly 45% of the sweets in a bag means the same as $\frac{45}{100}$ of them and $\frac{45}{100} = \frac{9}{20}$

i.e. $45\% = \frac{9}{20}$

EXERCISE 5F

Ninety per cent of the students who sat a test passed. What fraction was this?

$$90\% = \frac{90}{100}$$

$$= \frac{9}{10}$$

For each question express the given percentage as a fraction in its lowest terms.

1 Last summer 60% of the pupils in my class went on holiday.

2 Jim took a lot of geranium cuttings and had an 80% success rate.

3 At my youth club only 35% of the members are boys.

4 The rate of inflation is about 5%.

5 Only 15% of the people who have played for the club have scored a goal.

6 The Post Office claims that 95% of the letters posted by first class post arrive the following day.

7 A shoe inspection showed that 48% of the pupils were wearing brown shoes.

8 When value added tax was first introduced the rate was 8%.

9 In a cricket match Peter Jones scored 68% of all the runs scored by the team.

10 A survey showed that 32% of the pupils in year 7 needed to wear glasses.

Three-fifths of the spectators at a match were women. What percentage is this?

$$\frac{3}{5} = \frac{3 \times 20}{5 \times 20}$$

$$= \frac{60}{100}$$

$$= 60\%$$

> We need to express $\frac{3}{5}$ as a number of hundredths. We can do this by making $\frac{3}{5}$ into an equivalent fraction with denominator 100.

In each question from **11** to **15** express the fraction as a percentage.

11 Recently, at the local garage, $\frac{1}{5}$ of the cars tested failed to get an MOT test certificate.

12 At a youth club $\frac{17}{20}$ of those present took part in at least one sporting activity.

13 About $\frac{17}{50}$ of first year pupils see more than 20 hours of television a week.

14 Approximately $\frac{3}{5}$ of fifth form pupils have a Saturday job.

15 In one day $\frac{24}{25}$ of the components produced by a machine were satisfactory.

CHANGING BETWEEN PERCENTAGES AND DECIMALS

Consider 20% of the cars in a car-park again.

20% of the cars $= \dfrac{20}{100}$ of the cars

But $\qquad \dfrac{20}{100} = 0.20$

i.e. $\qquad 20\% = 0.20 = 0.2$

EXERCISE 5G

> Gail gave 46% of her tapes to her sister. What decimal part of the tapes did she give to her sister?
>
> $$46\% = \frac{46}{100}$$
> $$= 0.46$$

For each statement express the given percentage as a decimal.

1 At the beginning of term 65% of the pupils in the class needed a new exercise book.

2 When all the textbooks had been handed in it was discovered that 15% of them needed replacing.

3 A survey of the whole school revealed that 45% of the pupils had no brothers or sisters.

4 The police stopped all the vehicles on an approach road to a motorway and found that 6% of these vehicles had defective tyres.

5 A travel agent reported that 37% of all the holidays they arranged were to Spain.

6 In the school orchestra about 55% of the members play stringed instruments.

7 A consignment of oranges was delayed at the docks. As a result 24% of the oranges had to be destroyed.

8 A double glazing firm claimed that if their windows were installed there would be a 12% reduction in Mr Symes' heating bills.

> A close inspection of the components produced by the night shift showed that 0.12 of them were faulty. What percentage is this?
>
> $$0.12 = 12\%$$
>
> 0.12 means 12 hundredths, and 12 out of a hundred is 12%, so 0.12 = 12%.

In statements **9** to **12** express each decimal as a percentage.

9 At Berkley it rained on 0.33 of the days in the year.

10 A large number of new car buyers were interviewed. Red was the favourite colour chosen by 0.28 of them.

11 An opinion poll suggested that 0.78 of the population have a good idea now as to how they will vote in the next general election.

12 When a large number of electric light bulbs were tested, 0.92 of them burned for more than 100 hours.

INTERCHANGING DECIMALS AND FRACTIONS

EXERCISE 5H

Write as a fraction or as a mixed number in its lowest terms

a 0.6   **b** 12.04

a   units   tenths

$0.6 = 0 . 6 = \frac{6}{10}$

$= \frac{3}{5}$

b   tens   units   tenths   hundredths

$12.04 = 1 \quad 2 . \quad 0 \quad\quad 4 = 12\frac{4}{100}$

$= 12\frac{1}{25}$

In questions **1** to **8** write each decimal as a fraction in its lowest terms, using mixed numbers where necessary.

1 0.2   **3** 0.001   **5** 1.8   **7** 15.5

2 0.06   **4** 1.7   **6** 0.007   **8** 2.01

In each statement from **9** to **12** write the decimal as a fraction in its lowest terms.

9 It is estimated that 0.86 of the families in Northgate Street own a car.

10 In a wood 0.68 of the trees lose their leaves in the winter.

11 George observed that 0.48 of the cars that passed him carried only the driver.

12 There were 360 seats on the aircraft and only 0.05 of them were vacant.

> At a political meeting $\frac{3}{5}$ of the audience left before the end. Express this fraction as a decimal.
>
> $$\frac{3}{5} = \frac{3 \times 2}{5 \times 2}$$
>
> > We need to write $\frac{3}{5}$ as an equivalent fraction with denominator 10 or 100 or 1000, then we can write it as a decimal.
>
> $$= \frac{6}{10}$$
>
> $$= 0.6$$

In questions **13** to **23** express as a decimal.

13 $\frac{13}{100}$ **15** $\frac{4}{10}$ **17** $\frac{3}{20}$ **19** $\frac{23}{50}$

14 $\frac{7}{20}$ **16** $4\frac{7}{10}$ **18** $12\frac{1}{2}$ **20** $7\frac{9}{10}$

21 At a wedding reception $\frac{2}{5}$ of the guests did not drink sherry.

22 For an afternoon performance of Cinderella $\frac{13}{20}$ of the audience were children.

23 In a multi-storey car-park $\frac{3}{8}$ of the parked cars were at ground level.

EXPRESSING PARTS OF A WHOLE IN DIFFERENT WAYS

Now we can express a part of a whole either as a fraction, a decimal or as a percentage.

A bookseller is offering a book you would like to have at 15% below the list price. You would like to know by what fraction the price of the book is reduced.

To find this fraction you must convert 15% into a fraction,

i.e. $15\% = \frac{15}{100} = \frac{3}{20}$

The price of the book is being reduced by $\frac{3}{20}$.

EXERCISE 5I

> Express $\frac{4}{5}$ of the people at a concert as the percentage of people at the concert.
>
> $$\frac{4}{5} = \frac{4 \times 20}{5 \times 20}$$
> $$= \frac{80}{100}$$
> $$= 80\%$$
>
> i.e. $\frac{4}{5}$ of the people at a concert is the same as 80% of the people at the concert.

1 At a dancing class $\frac{13}{20}$ of those present were boys.

   a What percentage is this?

   b Express the fraction as a decimal.

2 When Hetty went to a car boot sale she found that $\frac{3}{10}$ of the vehicles were vans.
What percentage was this?

3 The registers showed that only 0.05 of the pupils in the first year had 100% attendance last term.

   a What fraction is this?

   b What percentage of the first year pupils had a 100% attendance last term?

4 Jo is given an 8% wage increase.
By what fraction will her wage increase?

5 Marion spends $\frac{21}{50}$ of her income on food and lodgings.

   a What percentage is this?

   b As a decimal, what part of her total income does she spend on food and lodging?

6 Marmalade consists of 28% fruit, $\frac{29}{50}$ sugar and the remainder water.

   a What fraction of the marmalade is fruit?

   b What percentage of the marmalade is sugar?

   c What percentage is water?

7 The value of a house increased by 90% over a 10-year period.
By what fraction did the value of the house increase?

8

One tablespoon of orange concentrate is added to 9 tablespoons of water to make a drink.

a What fraction of the drink is water?

b What percentage of the drink is concentrate?

(When we mix two quantities together to make a whole, we can use the word 'ratio' to describe how much of each we use. In this case concentrate and water are mixed in the ratio 1 to 9. We write this 1 : 9.)

9 An alloy is 60% copper, $\frac{1}{5}$ is nickel and the remainder is tin.

a What fraction is copper?

b What percentage is **i** nickel   **ii** either nickel or copper?

c Express the part that is tin as a decimal.

d What is the ratio of the amount of copper to the amount of tin?

EXERCISE 5J

1 Express as a fraction in its lowest terms

   a 30%   **b** 85%   **c** 42%   **d** 5%

2 Express as a decimal

   a 44%   **b** 68%   **c** 170%   **d** 16%

3 Express as a percentage

   a $\frac{2}{5}$ **b** $\frac{17}{20}$ **c** $\frac{1}{4}$ **d** $\frac{17}{25}$

4 Express as a percentage

   a 0.2   **b** 0.62   **c** 0.84   **d** 0.78

Copy and complete the following table.

	Fraction	Percentage	Decimal
5	$\frac{3}{4}$	75%	0.75
6	$\frac{4}{5}$		
7		60%	
8			0.7
9	$\frac{11}{20}$		
10		44%	

FRACTIONS OF A QUANTITY

Margot was given £20 in cash for her birthday. She was persuaded to put $\frac{3}{5}$ of it into a savings account. How much should she put in?

To find out how much she is to put in a savings account Margot needs to know that $\frac{3}{5}$ means 3 parts out of 5 equal parts.

To find $\frac{3}{5}$ of £20 we first divide 20 by 5 to find the size of one of the equal parts.

$$20 \div 5 = 4 \quad \text{so} \quad \tfrac{1}{5} \text{ of } £20 = £4$$

then
$$\tfrac{3}{5} \text{ of } £20 = £4 \times 3 = £12$$

EXERCISE 5K

Find $\frac{3}{5}$ of 95 metres.

$$\tfrac{1}{5} \text{ of } 95 = 95 \div 5 = 19$$

$\therefore \qquad \frac{3}{5}$ of $95 = 3 \times 19 = 57$

i.e. $\qquad \frac{3}{5}$ of 95 metres is 57 metres

Find $\frac{3}{4}$ of £1.

$$£1 = 100 \text{ pence}$$
$$\tfrac{1}{4} \text{ of } 100 = 100 \div 4 = 25$$

$\therefore \qquad \frac{3}{4}$ of $100 = 25 \times 3 = 75$

i.e. $\qquad \frac{3}{4}$ of £1 is 75 pence

Find

1 $\frac{3}{5}$ of 45 p

2 $\frac{9}{10}$ of 50 litres

3 $\frac{3}{8}$ of 88 miles

4 $\frac{7}{16}$ of 48 gallons

5 $\frac{4}{5}$ of 1 year of 365 days

6 $\frac{5}{8}$ of 16 dollars

7 $\frac{3}{7}$ of 35 miles

8 $\frac{4}{9}$ of 63 kilometres

9 $\frac{1}{7}$ of 1 week

10 $\frac{7}{8}$ of 1 day (24 hours)

11 In an election for club secretary Peter got $\frac{5}{12}$ of the votes and Sue $\frac{2}{5}$. If 60 people were entitled to vote,

   a how many voted for **i** Peter   **ii** Sue?

   b how many failed to vote?

12 Sally has a 60 metre ball of string. She used $\frac{3}{5}$ of it on Monday and $\frac{1}{4}$ of it on Tuesday.

   a What length of string did she use
     i on Monday   **ii** on Tuesday?

   b What length remained?

We can find the percentages of quantities in a similar way.

A joint of meat, weighing 1600 grams, contains 15% fat. What weight of fat is there in the joint?

The weight of fat is 15% of 1600 g

But $15\% = \frac{15}{100}$

        $\frac{1}{100}$ of $1600 = 1600 \div 100 = 16$

so $\frac{15}{100}$ of $1600 = 16 \times 15 = 240\,$g

$\therefore$ the amount of fat in the meat is 240 g

Find the value of

13 40% of £800

14 15% of 100 metres

15 33% of 600 kilograms

16 74% of 200 litres

17 40% of £48

18 6% of 3500 millimetres

19 30% of 500 litres

20 86% of 1100 grams

21 30% of £4200

22 25% of £60

23 Tim gets £12 a week and saves 25% of it.

    a How much does he save?   **b** How much does he spend?

24 In a choir 55% of the 80 choristers wear spectacles.
How many choristers

    a wear spectacles   **b** do not wear spectacles?

25 The audience for a performance of Oklahoma totalled 840. $\frac{1}{10}$ of
the audience were men, 45% were women and the remainder were
children.

    a How many in the audience were   **i** men   **ii** women?
    b How many children were there?

26 There are 250 first year pupils in a school. 26% of these pupils walk
to school, $\frac{3}{5}$ of them come by bus and the remainder are brought
by car. How many pupils

    a walk to school   **b** come by bus   **c** come by car?

**COMPARING THE
SIZES OF
FRACTIONS**

Alec has one box of light chocolates and one box of dark chocolates.
Each box contains the same number of chocolates. He gives $\frac{5}{7}$ of the
light chocolates to Bernard and $\frac{2}{3}$ of the dark chocolates to Alf. Does
he give more chocolates to Bernard than to Alf, or does Alf get the
greater number?

Before we can compare these two fractions we must change them into
the *same kind* of fraction. That means we must find equivalent fractions
that have the same denominator. This denominator must be a number
that both 7 and 3 divide into. So our new denominator is 21. Now

$$\frac{5}{7} = \frac{15}{21} \quad \text{and} \quad \frac{2}{3} = \frac{14}{21}$$

We can see that $\frac{15}{21}$ is bigger than $\frac{14}{21}$, i.e. $\frac{5}{7}$ is bigger than $\frac{2}{3}$,
so Bernard gets more chocolates.

We often use the symbol $>$ instead of writing 'is bigger than'.
Using this symbol we could write

$$\frac{15}{21} > \frac{14}{21}, \quad \text{so} \quad \frac{5}{7} > \frac{2}{3}.$$

Similarly we use $<$ instead of writing 'is less than'.

EXERCISE 5L

Which is the bigger fraction, $\frac{3}{5}$ or $\frac{7}{11}$?

$\frac{3}{5} = \frac{33}{55}$ and $\frac{7}{11} = \frac{35}{55}$ | 55 divides by 5 and by 11. |

So $\frac{7}{11}$ is the bigger fraction.

In the following questions find which is the bigger fraction.

1 $\frac{1}{2}$ or $\frac{1}{3}$ **3** $\frac{2}{3}$ or $\frac{4}{5}$ **5** $\frac{2}{7}$ or $\frac{3}{8}$

2 $\frac{3}{4}$ or $\frac{5}{6}$ **4** $\frac{2}{9}$ or $\frac{1}{7}$ **6** $\frac{2}{3}$ or $\frac{3}{4}$

7 $\frac{2}{5}$ of a field or $\frac{3}{7}$ of the field.

8 $\frac{4}{5}$ of an audience or $\frac{6}{7}$ of the audience.

9 $\frac{4}{7}$ of a bag of potatoes or $\frac{3}{5}$ of the bag.

10 $\frac{2}{9}$ of the cars sold or $\frac{3}{11}$ of them.

In questions **11** to **20**, put either $>$ or $<$ between the fractions.

11 $\frac{1}{4}$ $\frac{2}{7}$ **16** $\frac{8}{11}$ $\frac{3}{4}$

12 $\frac{2}{3}$ $\frac{5}{6}$ **17** $\frac{3}{5}$ $\frac{2}{3}$

13 $\frac{5}{8}$ $\frac{7}{10}$ **18** $\frac{2}{9}$ $\frac{3}{7}$

14 $\frac{1}{3}$ $\frac{2}{5}$ **19** $\frac{5}{8}$ $\frac{4}{5}$

15 $\frac{2}{9}$ $\frac{1}{5}$ **20** $\frac{1}{3}$ $\frac{2}{7}$

Arrange the following fractions in order of size with the smallest first.

$$\frac{3}{4}, \ \frac{7}{10}, \ \frac{1}{2}, \ \frac{4}{5}$$

$$\frac{3}{4} = \frac{15}{20}$$

Since 20 is the lowest number that denominators 4, 10, 2 and 5 will all go into exactly, we express every fraction in twentieths.

$$\frac{7}{10} = \frac{14}{20}$$

$$\frac{1}{2} = \frac{10}{20}$$

$$\frac{4}{5} = \frac{16}{20}$$

In ascending order the fractions are $\frac{1}{2}, \ \frac{7}{10}, \ \frac{3}{4}, \ \frac{4}{5}$

Arrange the following fractions in order of size with the smallest first.

21 $\frac{2}{3}, \ \frac{1}{2}, \ \frac{3}{5}, \ \frac{7}{10}$

22 $\frac{13}{20}, \ \frac{3}{4}, \ \frac{4}{10}, \ \frac{5}{8}$

23 $\frac{2}{5}, \ \frac{3}{8}, \ \frac{17}{20}, \ \frac{1}{2}, \ \frac{7}{10}$

24 The lengths of five small panel pins are

$\frac{7}{10}$ inch, $\frac{2}{5}$ inch, $\frac{3}{5}$ inch, $\frac{14}{25}$ inch, and $\frac{1}{2}$ inch.

Arrange these pins in order of size with the smallest first.

Arrange the following fractions in order of size with the largest first.

25 $\frac{5}{6}, \ \frac{1}{2}, \ \frac{7}{9}, \ \frac{11}{18}, \ \frac{2}{3}$

26 $\frac{13}{20}, \ \frac{3}{5}, \ \frac{1}{2}, \ \frac{3}{4}, \ \frac{7}{10}$

27 Mike has 5 drills. Their sizes are

$\frac{7}{16}$ inch, $\frac{1}{2}$ inch, $\frac{5}{8}$ inch, $\frac{19}{32}$ inch, and $\frac{3}{4}$ inch.

Arrange these drills in order of size with the smallest first.

COMPARING THE
SIZES OF
DECIMALS

To decide which of the decimals 4.67 or 4.63, is the larger

look first at the number of units – they are the same
next look at the number of tenths – they are also the same
finally look at the number of hundredths – 7 is bigger than 3 so **4.67**
is bigger than **4.63**

EXERCISE 5M

> Which is the larger 1.5 or 1.55?
>
> To compare these decimals write 1.5 as 1.50 showing that
> 1.55 is larger than 1.5

Which is the larger?

1 3.57 or 3.59   **2** 25.64 or 25.46   **3** 5 or 4.88

Put these numbers in ascending order.

4 6.76, 4.83, 6.29   **6** 9.07, 9.51, 9.18, 9.03

5 12.6, 14.09, 12.55, 13.75   **7** 7.555, 7.5, 7.05, 7.55

Put these numbers in descending order.

8 4.07, 4.87, 4.76, 4.69   **9** 9.88, 10.1, 7.06, 10.05

It is often easier to put fractions in order of size by converting them into decimals.

> Convert the fractions $\frac{3}{4}$, $\frac{4}{5}$, $\frac{17}{20}$, $\frac{7}{10}$, $\frac{17}{25}$ into decimals and
> use your answers to put the fractions in ascending order.
>
> $$\frac{3}{4} = 0.75$$
>
> $$\frac{4}{5} = 0.8$$
>
> $$\frac{17}{20} = 0.85$$
>
> $$\frac{7}{10} = 0.7$$
>
> $$\frac{17}{25} = 0.68$$
>
> In ascending order the decimals are 0.68, 0.7, 0.75, 0.8 and 0.85
> Therefore the fractions in ascending order are
> $\frac{17}{25}$, $\frac{7}{10}$, $\frac{3}{4}$, $\frac{4}{5}$ and $\frac{17}{20}$.

Use decimals to put the following fractions in ascending order.

10 $\frac{3}{5}$, $\frac{6}{25}$, $\frac{7}{10}$, $\frac{13}{20}$, $\frac{3}{4}$

11 $\frac{1}{2}$, $\frac{11}{25}$, $\frac{11}{20}$, $\frac{4}{5}$, $\frac{9}{10}$

12 The thicknesses of four sheets of metal are $\frac{35}{1000}$ inch, $\frac{13}{100}$ inch, $\frac{97}{1000}$ inch and $\frac{1}{10}$ inch. Arrange these sheets in ascending order of thickness.

MIXED EXERCISE

EXERCISE 5N

1 Fill in the missing numbers to make equivalent fractions

    a $\frac{5}{7} = \frac{}{28}$ **b** $\frac{7}{15} = \frac{}{60}$ **c** $\frac{19}{20} = \frac{57}{}$ **d** $\frac{5}{12} = \frac{}{144}$

2

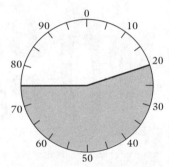

Part of this circle is shaded. Express the shaded part as

    a a percentage   **b** a decimal   **c** a fraction in its lowest terms.

3 Express

    a 65% as a fraction in its lowest terms

    b 78% as a decimal

    c $\frac{27}{50}$ as a percentage

    d $\frac{19}{20}$ as a decimal

    e 0.41 as a percentage

    f 0.44 as a fraction in its lowest terms.

4 Find   **a** $\frac{7}{12}$ of 84 m   **b** 35% of 440 kg   **c** 0.76 of 550 cm

5 Which is the larger   **a** 2.54 or 2.6   **b** $\frac{5}{7}$ or $\frac{7}{10}$?

6 The thicknesses of four sheets of plastic are

$\frac{7}{20}$ mm, $\frac{1}{2}$ mm, $\frac{31}{50}$ mm and $\frac{4}{10}$ mm.

By converting these fractions into decimals, or otherwise, put them in ascending order.

PRACTICAL WORK

You need twelve 2 p coins for this work, together with an exact copy of the outside edge of the rectangular ticket shown above.
Place the coins on the rectangle so that they all fit on it. The coins should touch one another without overlapping and the outer coins of the arrangement must touch the edges of the ticket.

1 How many coins touch the edges of the ticket?
What fraction of the total number of coins is this?

2 How many coins touch exactly two other coins?
What fraction of the total number of coins is this?

3 What fraction of the total number of coins touch

   a 3 other coins   **b** 4 other coins?

(All answers in fractions should be given in their lowest terms.)

INVESTIGATION

Using the numbers 1, 2, 4 and 8 write down all the fractions you can think of that are equal to or smaller than 1. Use a single number for the numerator and a single number for the denominator. You are allowed to use a number more than once in the same fraction.

a **i** How many different fractions can you find?

  ii Which fractions are equivalent fractions?

  iii What fraction does not have an equivalent fraction in the list you have written?

b Add 16 to the list of numbers 1, 2, 4, 8, and repeat part **a**.

c Two-digit numbers, such as 14 and 82, can be made from the digits 1, 2, 4 and 8. Use such numbers to repeat part **a**.

ADDITION AND SUBTRACTION OF FRACTIONS AND DECIMALS

6

We know that part of a whole can be expressed as a fraction, as a decimal or as a percentage. If we use fractions we must be able to add and subtract them.

Class 7W plan to raise some money for charity and decide how they will divide the sum they collect. The pupils decide to give $\frac{1}{4}$ of the sum collected to the NSPCC, $\frac{1}{3}$ to the local hospice and the remainder to OXFAM.

Anna commented 'That leaves more than half for OXFAM!' Hasib quickly replied 'No, it doesn't.'

Do you know who is right?

Sam joined in the argument with 'But $\frac{1}{3}$ is more than $\frac{1}{4}$, so the two together must be more than $\frac{1}{2}$.'

Is this a sensible statement?

EXERCISE 6A

1 Think of problems where fractions are involved.

2 Discuss what you need to be able to do with fractions to solve these problems.

In an audience $\frac{1}{15}$ are men, $\frac{7}{15}$ are women, $\frac{1}{5}$ are boys and the remainder are girls.

To find out what fraction are males we need to be able to add $\frac{1}{15}$ and $\frac{1}{5}$, and to find out what fraction are girls we must be able to add $\frac{1}{15}$, $\frac{7}{15}$ and $\frac{1}{5}$, and subtract the total from 1.

This example, together with your discussions should convince you of the need to be able to add and subtract fractions, and to have an idea whether your answers are roughly correct.

ADDING FRACTIONS

Suppose there is a bowl of oranges and apples. First you take three oranges and then two more oranges. You then have five oranges; we can add the 3 and the 2 together because they are the same kind of fruit. But three oranges and two apples cannot be added together because they are different kinds of fruit.

For fractions it is the denominator that tells us the kind of fraction, so we can add fractions together if they have the same denominator but not while their denominators are different.

EXERCISE 6B

Add the fractions

a $\dfrac{2}{7}+\dfrac{3}{7}$ **b** $\dfrac{9}{22}+\dfrac{5}{22}$, simplifying the answers where you can.

a $\quad\dfrac{2}{7}+\dfrac{3}{7}=\dfrac{2+3}{7}$

$\qquad\quad=\dfrac{5}{7}$

b $\quad\dfrac{9}{22}+\dfrac{5}{22}=\dfrac{9+5}{22}$

$\qquad\quad=\dfrac{\cancel{14}^{7}}{\cancel{22}_{11}}$

We divide the top and bottom by 2 to get a simpler fraction.

$\qquad\quad=\dfrac{7}{11}$

Add the fractions given in questions **1** to **18**, simplifying the answers where you can.

1 $\dfrac{1}{4}+\dfrac{2}{4}$ **7** $\dfrac{11}{23}+\dfrac{8}{23}$ **13** $\dfrac{2}{21}+\dfrac{9}{21}$

2 $\dfrac{1}{8}+\dfrac{3}{8}$ **8** $\dfrac{1}{7}+\dfrac{2}{7}$ **14** $\dfrac{7}{30}+\dfrac{8}{30}$

3 $\dfrac{3}{11}+\dfrac{2}{11}$ **9** $\dfrac{2}{5}+\dfrac{1}{5}$ **15** $\dfrac{6}{13}+\dfrac{5}{13}$

4 $\dfrac{3}{13}+\dfrac{7}{13}$ **10** $\dfrac{3}{10}+\dfrac{1}{10}$ **16** $\dfrac{1}{10}+\dfrac{7}{10}$

5 $\dfrac{2}{7}+\dfrac{4}{7}$ **11** $\dfrac{5}{16}+\dfrac{7}{16}$ **17** $\dfrac{4}{11}+\dfrac{2}{11}$

6 $\dfrac{8}{30}+\dfrac{19}{30}$ **12** $\dfrac{21}{100}+\dfrac{19}{100}$ **18** $\dfrac{7}{15}+\dfrac{3}{15}$

We can add more than two fractions in the same way.

Add the fractions $\dfrac{3}{17} + \dfrac{5}{17} + \dfrac{8}{17}$

$$\frac{3}{17} + \frac{5}{17} + \frac{8}{17} = \frac{3+5+8}{17}$$
$$= \frac{16}{17}$$

Add the fractions given in questions **19** to **24**.

19 $\dfrac{2}{15} + \dfrac{4}{15} + \dfrac{6}{15}$

20 $\dfrac{8}{100} + \dfrac{21}{100} + \dfrac{11}{100}$

21 $\dfrac{4}{45} + \dfrac{11}{45} + \dfrac{8}{45} + \dfrac{2}{45}$

22 $\dfrac{1}{14} + \dfrac{3}{14} + \dfrac{5}{14} + \dfrac{2}{14}$

23 $\dfrac{2}{51} + \dfrac{4}{51} + \dfrac{6}{51} + \dfrac{8}{51} + \dfrac{7}{51}$

24 $\dfrac{3}{99} + \dfrac{11}{99} + \dfrac{4}{99} + \dfrac{7}{99}$

25 If $\dfrac{5}{8}$ of the spectators at a match are men and $\dfrac{1}{8}$ are women, what fraction of the spectators are adults?

26 In a shop $\dfrac{1}{10}$ of the videos are films, $\dfrac{3}{10}$ sport and $\dfrac{3}{10}$ are documentaries.
What fraction of the videos are either films, sport or documentaries?

FRACTIONS WITH DIFFERENT DENOMINATORS

To add fractions with different denominators we must first change the fractions into equivalent fractions with the same denominator. This new denominator must be a number that both original denominators divide into.

For instance, if we want to add $\dfrac{2}{5}$ and $\dfrac{3}{7}$ we choose 35 for our new denominator because 35 can be divided by both 5 and 7:

$$\frac{2}{5} = \frac{14}{35}$$

$$\frac{3}{7} = \frac{15}{35}$$

So $\qquad \dfrac{2}{5} + \dfrac{3}{7} = \dfrac{14}{35} + \dfrac{15}{35} = \dfrac{29}{35}$

EXERCISE 6C

Find $\dfrac{2}{7} + \dfrac{3}{8}$

$\dfrac{2}{7} + \dfrac{3}{8} = \dfrac{16}{56} + \dfrac{21}{56} = \dfrac{37}{56}$

> 7 and 8 both divide into 56, so we use 56 as the new denominator.

Find

1 $\dfrac{2}{3} + \dfrac{1}{5}$

5 $\dfrac{3}{10} + \dfrac{2}{3}$

9 $\dfrac{1}{6} + \dfrac{2}{7}$

2 $\dfrac{1}{5} + \dfrac{3}{8}$

6 $\dfrac{4}{7} + \dfrac{1}{8}$

10 $\dfrac{5}{6} + \dfrac{1}{7}$

3 $\dfrac{1}{5} + \dfrac{1}{6}$

7 $\dfrac{3}{7} + \dfrac{1}{6}$

11 $\dfrac{3}{11} + \dfrac{5}{9}$

4 $\dfrac{2}{5} + \dfrac{3}{7}$

8 $\dfrac{2}{3} + \dfrac{2}{7}$

12 $\dfrac{2}{9} + \dfrac{3}{10}$

The new denominator, which is called the *common denominator*, is not always as big as you might first think. For instance, if we want to add $\dfrac{3}{4}$ and $\dfrac{1}{12}$, the common denominator is 12 because it can be divided exactly by 4 and by 12.

Find $\dfrac{3}{4} + \dfrac{1}{12}$

$\dfrac{3}{4} + \dfrac{1}{12} = \dfrac{9}{12} + \dfrac{1}{12}$

$= \dfrac{10^5}{12^6}$

$= \dfrac{5}{6}$

Find

13 $\dfrac{2}{5} + \dfrac{3}{10}$

17 $\dfrac{1}{5} + \dfrac{7}{10}$

21 $\dfrac{1}{20} + \dfrac{3}{5}$

14 $\dfrac{3}{8} + \dfrac{7}{16}$

18 $\dfrac{1}{4} + \dfrac{3}{8}$

22 $\dfrac{4}{11} + \dfrac{5}{22}$

15 $\dfrac{3}{7} + \dfrac{8}{21}$

19 $\dfrac{2}{3} + \dfrac{2}{9}$

23 $\dfrac{2}{5} + \dfrac{7}{15}$

16 $\dfrac{3}{10} + \dfrac{3}{100}$

20 $\dfrac{4}{9} + \dfrac{5}{18}$

24 $\dfrac{7}{12} + \dfrac{1}{6}$

More than two fractions can be added in this way. The common denominator must be divisible by *all* of the original denominators.

Find $\dfrac{1}{8} + \dfrac{1}{2} + \dfrac{1}{3}$

$$\dfrac{1}{8} + \dfrac{1}{2} + \dfrac{1}{3} = \dfrac{3}{24} + \dfrac{12}{24} + \dfrac{8}{24}$$

8, 2 and 3 all divide into 24 so 24 is the new denominator.

$$= \dfrac{3 + 12 + 8}{24}$$

$$= \dfrac{23}{24}$$

Find

25 $\dfrac{1}{5} + \dfrac{1}{4} + \dfrac{1}{2}$

28 $\dfrac{5}{12} + \dfrac{1}{6} + \dfrac{1}{3}$

31 $\dfrac{1}{3} + \dfrac{2}{9} + \dfrac{1}{6}$

26 $\dfrac{1}{8} + \dfrac{1}{4} + \dfrac{1}{3}$

29 $\dfrac{1}{3} + \dfrac{1}{6} + \dfrac{1}{2}$

32 $\dfrac{2}{15} + \dfrac{1}{10} + \dfrac{2}{5}$

27 $\dfrac{3}{10} + \dfrac{2}{5} + \dfrac{1}{4}$

30 $\dfrac{1}{2} + \dfrac{3}{8} + \dfrac{1}{10}$

33 $\dfrac{1}{4} + \dfrac{1}{12} + \dfrac{1}{3}$

34 In a car-park $\dfrac{3}{7}$ of the cars were manufactured in the UK and $\dfrac{5}{14}$ in mainland Europe.
What fraction were made in Europe?

35 In a class $\dfrac{1}{5}$ of the pupils are less than 120 cm in height and $\dfrac{3}{10}$ are between 120 cm and 130 cm. No pupil is exactly 120 cm or 130 cm.
What fraction of the class is less than 130 cm?

36 When Kirsty comes home from shopping $\dfrac{1}{12}$ of the weight in her basket is green vegetables, $\dfrac{1}{3}$ is fruit and $\dfrac{1}{4}$ is other root vegetables.
What fraction of the weight she carries is fruit and vegetables?

37 Decide, without adding the fractions, whether $\dfrac{1}{4} + \dfrac{1}{5}$ is more than or less than $\dfrac{1}{2}$.
Give reasons for your answer.

38 Three possible answers are given for each 'sum'. Two of them are obviously wrong. Without working out the sum, decide which answer is most likely to be correct.

a $\frac{1}{5} + \frac{1}{7} = \ldots$ **A** $\frac{1}{4}$ **B** $\frac{1}{35}$ **C** $\frac{12}{35}$

b $\frac{2}{7} + \frac{1}{3} = \ldots$ **A** $\frac{3}{10}$ **B** $\frac{13}{21}$ **C** $\frac{2}{21}$

c $\frac{5}{18} + \frac{2}{9} = \ldots$ **A** $\frac{1}{2}$ **B** $\frac{10}{27}$ **C** $\frac{7}{27}$

39 Lyn adds two fractions and writes down

$$\frac{1}{4} + \frac{2}{5} = \frac{5}{20} + \frac{4}{20} = \frac{9}{40}$$

How many mistakes are there? Explain why they are mistakes.

SUBTRACTING FRACTIONS

Exactly the same method is used for subtracting fractions as for adding them. To work out the value of $\frac{7}{8} - \frac{3}{8}$ we notice that the denominators are the same, so

$$\frac{7}{8} - \frac{3}{8} = \frac{7 - 3}{8}$$

$$= \frac{\cancel{4}^{1}}{\cancel{8}^{2}}$$

$$= \frac{1}{2}$$

EXERCISE 6D

Find $\frac{7}{9} - \frac{1}{4}$

$$\frac{7}{9} - \frac{1}{4} = \frac{28}{36} - \frac{9}{36}$$

$$= \frac{28 - 9}{36}$$

$$= \frac{19}{36}$$

> The denominators are not the same so we use equivalent fractions with denominator 36.

Find

1 $\frac{8}{9} - \frac{2}{9}$ **3** $\frac{3}{4} - \frac{1}{5}$ **5** $\frac{2}{3} - \frac{3}{7}$

2 $\frac{7}{10} - \frac{2}{10}$ **4** $\frac{9}{10} - \frac{1}{2}$ **6** $\frac{11}{15} - \frac{4}{15}$

7 $\frac{8}{11} - \frac{2}{5}$

11 $\frac{19}{100} - \frac{1}{10}$

15 $\frac{3}{4} - \frac{5}{8}$

8 $\frac{7}{9} - \frac{2}{3}$

12 $\frac{5}{8} - \frac{2}{7}$

16 $\frac{7}{12} - \frac{1}{3}$

9 $\frac{8}{13} - \frac{1}{2}$

13 $\frac{15}{16} - \frac{3}{4}$

17 $\frac{13}{18} - \frac{5}{9}$

10 $\frac{11}{12} - \frac{5}{6}$

14 $\frac{7}{15} - \frac{1}{5}$

18 $\frac{13}{15} - \frac{3}{5}$

19 In the supermarket $\frac{7}{12}$ of Paula's bill is for fruit and vegetables. If $\frac{1}{8}$ is for fruit, what fraction is for vegetables?

20 Decide, without subtracting the fractions, if $\frac{3}{4} - \frac{1}{3}$ is less than or more than $\frac{1}{2}$.
Give reasons for your answer.

21 In a class $\frac{2}{5}$ of the pupils are less than 115 cm in height and $\frac{7}{10}$ are less than 130 cm. No pupil is exactly 115 cm.
What fraction of the class is more than 115 cm but less than 130 cm height?

22 Three possible answers are given for each of the following subtractions. Without working them out, choose the answer you think is most likely to be correct.

a $\frac{5}{9} - \frac{1}{3} = \ldots$ **A** $\frac{2}{9}$ **B** $\frac{5}{6}$ **C** $\frac{2}{3}$

b $\frac{14}{15} - \frac{4}{5} = \ldots$ **A** 1 **B** $\frac{2}{3}$ **C** $\frac{2}{15}$

c $\frac{19}{21} - \frac{3}{7} = \ldots$ **A** $\frac{16}{21}$ **B** $\frac{10}{21}$ **C** $\frac{4}{7}$

23 The following calculation contains at least one mistake.
Point out any mistake and give a reason why you think it is a mistake.

$$\frac{11}{12} - \frac{7}{9} = \frac{33}{36} - \frac{27}{36} = \frac{6}{36} = \frac{1}{3}$$

**ADDING AND
SUBTRACTING
FRACTIONS**

Fractions can be added and subtracted in one problem in a similar way. For example

$$\frac{7}{9} + \frac{1}{18} - \frac{1}{6} = \frac{14}{18} + \frac{1}{18} - \frac{3}{18}$$

$$= \frac{14 + 1 - 3}{18}$$

$$= \frac{12}{18}$$

$$= \frac{2}{3}$$

It is not always possible to work from left to right in order because we have to subtract too much too soon. In this case we can do the adding first. Remember that it is the sign *in front* of a number that tells you what to do with that number.

EXERCISE 6E

Find $\frac{1}{8} - \frac{3}{4} + \frac{11}{16}$

$$\frac{1}{8} - \frac{3}{4} + \frac{11}{16} = \frac{2}{16} - \frac{12}{16} + \frac{11}{16}$$

$$= \frac{13 - 12}{16}$$

$$= \frac{1}{16}$$

Find $2 + 11$ first.

Find

1 $\frac{3}{4} + \frac{1}{2} - \frac{7}{8}$

2 $\frac{6}{7} - \frac{9}{14} + \frac{1}{2}$

3 $\frac{3}{8} + \frac{7}{16} - \frac{3}{4}$

4 $\frac{3}{5} + \frac{3}{25} - \frac{27}{50}$

5 $\frac{2}{3} + \frac{1}{6} - \frac{5}{12}$

6 $\frac{4}{5} - \frac{7}{10} + \frac{1}{2}$

7 $\frac{7}{10} - \frac{41}{100} + \frac{1}{20}$

8 $\frac{5}{8} - \frac{21}{40} + \frac{2}{5}$

9 $\frac{7}{12} - \frac{1}{6} + \frac{1}{3}$

10 $\frac{2}{9} - \frac{1}{3} + \frac{1}{6}$

11 $\frac{1}{6} - \frac{2}{3} + \frac{7}{12}$

12 $\frac{1}{8} - \frac{13}{16} + \frac{3}{4}$

13 $\frac{1}{6} - \frac{5}{18} + \frac{1}{3}$

14 $\frac{1}{5} - \frac{7}{10} + \frac{17}{20}$

15 $\frac{2}{3} - \frac{5}{6} + \frac{1}{2}$

16 $\frac{3}{10} - \frac{61}{100} + \frac{1}{2}$

17 $\frac{1}{8} - \frac{7}{24} + \frac{5}{12}$

18 $\frac{3}{10} + \frac{2}{15} - \frac{2}{5}$

19 Some scientific calculators will work with fractions.

Using a Casio, $\frac{1}{2}+\frac{1}{3}$ can be found as a fraction by pressing

| 1 | $a\frac{b}{c}$ | 2 | + | 1 | $a\frac{b}{c}$ | 3 | = |

The display then reads $5\!\rfloor 6$; i.e. $\frac{1}{2}+\frac{1}{3}=\frac{5}{6}$

Use your calculator, if it will work in fractions, to check your answers to questions **1** to **18**.

In a class of school children, $\frac{1}{3}$ of the children come to school by bus, $\frac{1}{4}$ come to school on bicycles and the rest walk to school.

What fraction of the children ride to school? What fraction do not use a bus?

$$\text{The fraction who ride to school on bicycle and bus} = \frac{1}{3}+\frac{1}{4}$$
$$= \frac{4+3}{12}$$
$$= \frac{7}{12}$$

Therefore $\frac{7}{12}$ of the children ride to school.

The complete class of children is a whole unit, i.e. 1.

The fraction of children who do not use a bus is found by taking the bus users from the complete class, i.e.

$$\frac{1}{1}-\frac{1}{3}=\frac{3-1}{3}=\frac{2}{3}$$

---

20 A girl spends $\frac{1}{5}$ of her pocket money on sweets and $\frac{2}{3}$ on records.
What fraction has she spent?
What fraction has she left?

21 A group of friends went to a hamburger bar; $\frac{2}{5}$ of them bought a hamburger, $\frac{1}{3}$ of them just bought chips. The rest bought cola.
What fraction of the group bought food?
What fraction bought a drink?

22 At a pop festival, $\frac{2}{3}$ of the groups were all male, $\frac{1}{4}$ of the groups included one girl and the rest included more than one girl.
What fraction of the groups

    a were not all male   **b** contained more than one girl?

23 At a Youth Club, $\frac{1}{2}$ of the meetings are for playing table tennis, $\frac{1}{8}$ of the meetings are discussions and the rest are record sessions. What fraction of the meetings are

a record sessions   **b** not for discussions?

24 George Bentley had 3 sons and 1 daughter. When he died he left $\frac{3}{8}$ of his estate to his wife, $\frac{3}{32}$ to the youngest son, twice as much to his next son and three times as much to the eldest son as he had left to the youngest son. The remainder went to the daughter. What fraction did the daughter get?

25 Jim can dig a plot of ground in 5 hours and Ray can dig it in 6 hours.
What fraction of the plot can each of them dig in 1 hour?
If they work together, what fraction can they dig in 1 hour?
What fraction remains to be dug after 2 hours?

MIXED NUMBERS AND IMPROPER FRACTIONS

Most of the fractions we have met so far have been less than a whole unit. These are called *proper* fractions. But we often have more than a whole unit. Suppose, for instance, that we have one and a quarter bars of chocolate:

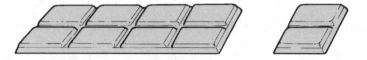

We have $1\frac{1}{4}$ bars, and $1\frac{1}{4}$ is called a mixed number.

Another way of describing the amount of chocolate is to say that we have five quarter bars.

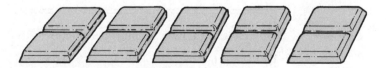

We have $\frac{5}{4}$ bars and $\frac{5}{4}$ is called an *improper* fraction because the numerator is bigger than the denominator.

But the amount of chocolate in the two examples is the same, so

$$\frac{5}{4} = 1\frac{1}{4}$$

Improper fractions can be changed into mixed numbers by finding out how many whole units there are. For instance, to change $\frac{8}{3}$ into a mixed number we look for the biggest number below 8 that divides by 3, i.e. 6. Then

$$\frac{8}{3} = \frac{6+2}{3} = \frac{6}{3} + \frac{2}{3} = 2 + \frac{2}{3} = 2\frac{2}{3}$$

We can also change mixed numbers into improper fractions. For instance, in $2\frac{4}{5}$ we have two whole units and $\frac{4}{5}$. In each whole unit there are five fifths, so in $2\frac{4}{5}$ we have ten fifths and four fifths, i.e.

$$2\frac{4}{5} = \frac{10}{5} + \frac{4}{5} = \frac{14}{5}$$

EXERCISE 6F

Change $\frac{15}{4}$ into a mixed number.

$\frac{15}{4} = \frac{12+3}{4}$

> 12 is the biggest number below 15 that divides exactly by 4.

$\quad = \frac{12}{4} + \frac{3}{4}$

$\quad = 3\frac{3}{4}$

In questions **1** to **15** change the improper fractions into mixed numbers.

1 $\frac{9}{4}$	**4** $\frac{53}{10}$	**7** $\frac{27}{4}$	**10** $\frac{114}{11}$	**13** $\frac{41}{3}$
2 $\frac{19}{4}$	**5** $\frac{88}{9}$	**8** $\frac{41}{8}$	**11** $\frac{83}{7}$	**14** $\frac{67}{5}$
3 $\frac{37}{6}$	**6** $\frac{7}{2}$	**9** $\frac{127}{5}$	**12** $\frac{91}{6}$	**15** $\frac{49}{10}$

Change $3\frac{1}{7}$ into an improper fraction.

$3\frac{1}{7} = 3 + \frac{1}{7}$

> There are 7 sevenths in 1, so there are 7×3, i.e. 21 sevenths in 3.

$\quad = \frac{21}{7} + \frac{1}{7}$

$\quad = \frac{22}{7}$

In questions **16** to **30** change the mixed numbers into improper fractions.

16 $4\frac{1}{3}$ **19** $10\frac{8}{9}$ **22** $2\frac{6}{7}$ **25** $5\frac{1}{2}$ **28** $8\frac{3}{4}$

17 $8\frac{1}{4}$ **20** $8\frac{1}{7}$ **23** $4\frac{1}{6}$ **26** $7\frac{2}{5}$ **29** $1\frac{9}{10}$

18 $1\frac{7}{10}$ **21** $6\frac{3}{5}$ **24** $3\frac{2}{3}$ **27** $3\frac{4}{5}$ **30** $7\frac{3}{8}$

THE MEANING OF
15 ÷ 4

$15 \div 4$ means 'how many fours are there in 15?'

There are 3 fours in 15 with 3 left over, so $15 \div 4 = 3$, remainder 3.

Now that remainder, 3, is $\frac{3}{4}$ of 4. Thus we can say that there are $3\frac{3}{4}$ fours in 15

i.e. $$15 \div 4 = 3\frac{3}{4}$$

But $$\frac{15}{4} = 3\frac{3}{4}$$

Therefore
$15 \div 4$ and $\frac{15}{4}$ mean the same thing

EXERCISE 6G

Calculate $27 \div 8$, giving your answer as a mixed number.

$$27 \div 8 = \frac{27}{8}$$

$$= 3\frac{3}{8}$$

Calculate the following divisions, giving your answers as mixed numbers.

1 $36 \div 7$ **5** $82 \div 5$ **9** $98 \div 12$

2 $59 \div 6$ **6** $29 \div 4$ **10** $107 \div 10$

3 $52 \div 11$ **7** $41 \div 3$ **11** $37 \div 5$

4 $20 \div 8$ **8** $64 \div 9$ **12** $52 \div 8$

ADDING MIXED NUMBERS

If we want to find the value of $2\frac{1}{3} + 3\frac{1}{4}$ we add the whole numbers and then the fractions, i.e.

$$2\frac{1}{3} + 3\frac{1}{4} = 2 + 3 + \frac{1}{3} + \frac{1}{4}$$

$$= 5 + \frac{4+3}{12}$$

$$= 5 + \frac{7}{12}$$

$$= 5\frac{7}{12}$$

Sometimes there is an extra step in the calculation. For example

$$3\frac{1}{2} + 2\frac{3}{8} + 5\frac{1}{4} = 3 + 2 + 5 + \frac{1}{2} + \frac{3}{8} + \frac{1}{4}$$

$$= 10 + \frac{4+3+2}{8}$$

$$= 10 + \frac{9}{8}$$

But $\frac{9}{8}$ is an improper fraction, so we change it into a mixed number

i.e.

$$3\frac{1}{2} + 2\frac{3}{8} + 5\frac{1}{4} = 10 + \frac{8+1}{8}$$

$$= 10 + 1 + \frac{1}{8}$$

$$= 11\frac{1}{8}$$

EXERCISE 6H

Find $2\frac{1}{2} + 4\frac{3}{4} + 1\frac{1}{6}$

$$2\frac{1}{2} + 4\frac{3}{4} + 1\frac{1}{6} = 7\frac{6+9+2}{12} \qquad \text{Add whole numbers.}$$

$$= 7\frac{17}{12} \qquad \text{Add fractions.}$$

$$= 8\frac{5}{12} \qquad \text{Split } \frac{17}{12} \text{ into } 1 + \frac{5}{12}.$$

Find

1 $2\frac{1}{4}+3\frac{1}{2}$

6 $1\frac{1}{3}+2\frac{5}{6}$

2 $1\frac{1}{2}+2\frac{1}{3}$

7 $3\frac{1}{4}+1\frac{1}{5}$

3 $4\frac{1}{5}+1\frac{3}{8}$

8 $2\frac{1}{7}+1\frac{1}{14}$

4 $5\frac{1}{9}+4\frac{1}{3}$

9 $6\frac{3}{10}+1\frac{2}{5}$

5 $3\frac{1}{4}+2\frac{5}{9}$

10 $8\frac{1}{7}+5\frac{2}{3}$

11 $7\frac{3}{8}+3\frac{7}{16}$

16 $2\frac{7}{10}+9\frac{1}{5}$

12 $1\frac{3}{4}+4\frac{7}{12}$

17 $5\frac{7}{10}+2\frac{3}{5}$

13 $3\frac{5}{7}+7\frac{1}{2}$

18 $9\frac{2}{3}+8\frac{5}{6}$

14 $6\frac{1}{2}+1\frac{9}{16}$

19 $2\frac{4}{5}+7\frac{3}{10}$

15 $8\frac{7}{8}+3\frac{3}{16}$

20 $6\frac{3}{10}+4\frac{4}{5}$

21 $1\frac{1}{4}+3\frac{2}{3}+6\frac{7}{12}$

23 $3\frac{3}{4}+5\frac{1}{8}+8\frac{5}{16}$

22 $5\frac{1}{7}+4\frac{1}{2}+7\frac{11}{14}$

24 $10\frac{2}{3}+3\frac{1}{6}+7\frac{2}{9}$

25 Sid wanted to raise his front gate because it was dragging on the ground when it opened. He put three washers on each support: one $\frac{1}{8}$ inch thick, one $\frac{3}{16}$ inch thick and one $\frac{1}{16}$ inch thick. How much higher did this make the gate?

26 George has three pieces of wood. One is $10\frac{1}{8}$ ft long, another $6\frac{1}{2}$ ft long and the third is $5\frac{7}{8}$ ft long.
How far do they stretch if they are laid end to end?

27 A recipe for sugar crisps requires $\frac{1}{2}$ cup fat, $\frac{3}{4}$ cup sugar, $\frac{2}{3}$ cup syrup and $2\frac{1}{2}$ cups flour.
How many cups of ingredients is this altogether?

28 Is $6\frac{1}{4}$ bigger or smaller than the sum of $3\frac{1}{3}$ and $2\frac{7}{8}$?

29 Hank has a length of timber that is $9\frac{1}{2}$ ft long. He wants to cut off three pieces, one $1\frac{5}{8}$ ft long, a second twice as long, and a third three times as long as the first.
Does he have enough wood?

SUBTRACTING MIXED NUMBERS

If we want to find the value of $5\frac{3}{4} - 2\frac{2}{5}$ we can use the same method as for adding.

$$5\frac{3}{4} - 2\frac{2}{5} = 5 - 2 + \frac{3}{4} - \frac{2}{5}$$

$$= 3 + \frac{15 - 8}{20}$$

$$= 3 + \frac{7}{20}$$

$$= 3\frac{7}{20}$$

But when we find the value of $6\frac{1}{4} - 2\frac{4}{5}$ we get

$$6\frac{1}{4} - 2\frac{4}{5} = 6 - 2 + \frac{1}{4} - \frac{4}{5}$$

$$= 4 + \frac{1}{4} - \frac{4}{5}$$

This time it is not so easy to deal with the fractions because $\frac{4}{5}$ is bigger than $\frac{1}{4}$. So we take one of the whole units and change it into a fraction, giving

$$3 + 1 + \frac{1}{4} - \frac{4}{5}$$

$$= 3 + \frac{20 + 5 - 16}{20}$$

$$= 3 + \frac{9}{20}$$

$$= 3\frac{9}{20}$$

EXERCISE 61   Find

1 $2\frac{3}{4} - 1\frac{1}{8}$ **5** $7\frac{3}{4} - 2\frac{1}{3}$ **9** $4\frac{4}{5} - 3\frac{1}{10}$

2 $3\frac{2}{3} - 1\frac{4}{5}$ **6** $3\frac{5}{6} - 2\frac{1}{3}$ **10** $6\frac{5}{7} - 3\frac{2}{5}$

3 $1\frac{5}{6} - \frac{2}{3}$ **7** $2\frac{6}{7} - 1\frac{1}{2}$ **11** $3\frac{1}{3} - 1\frac{1}{5}$

4 $3\frac{1}{4} - 2\frac{1}{2}$ **8** $4\frac{1}{2} - 2\frac{1}{5}$ **12** $5\frac{3}{4} - 2\frac{1}{2}$

13 $8\frac{4}{5} - 5\frac{1}{2}$ **17** $7\frac{1}{2} - 5\frac{3}{4}$ **21** $8\frac{6}{7} - 5\frac{3}{4}$

14 $5\frac{7}{9} - 3\frac{5}{7}$ **18** $4\frac{3}{5} - 1\frac{1}{4}$ **22** $3\frac{1}{2} - 1\frac{7}{8}$

15 $4\frac{5}{8} - 1\frac{1}{3}$ **19** $7\frac{6}{7} - 4\frac{3}{5}$ **23** $2\frac{1}{2} - 1\frac{3}{4}$

16 $6\frac{3}{4} - 3\frac{6}{7}$ **20** $8\frac{8}{11} - 2\frac{2}{3}$ **24** $5\frac{4}{7} - 3\frac{4}{5}$

25 $3\frac{1}{4} - 1\frac{7}{8}$ **27** $8\frac{2}{3} - 7\frac{8}{9}$ **29** $9\frac{7}{10} - 5\frac{4}{5}$

26 $5\frac{3}{5} - 2\frac{9}{10}$ **28** $4\frac{1}{6} - 2\frac{2}{3}$ **30** $2\frac{5}{12} - 1\frac{3}{4}$

31 Sheila cuts a piece of metal $5\frac{1}{2}$ inches long from a bar $27\frac{7}{10}$ inches long. What length remains?

32 A recipe for frozen Christmas puddings lists the main ingredients as: $1\frac{1}{2}$ cups of vanilla wafer crumbs, $\frac{1}{2}$ cup chopped nuts, $\frac{1}{2}$ cup chopped dates, $\frac{1}{2}$ cup chopped fruit peel, $\frac{1}{4}$ cup hot orange juice, $\frac{1}{3}$ cup sugar and 1 cup whipped cream.
If all the ingredients total $4\frac{3}{4}$ cups how many cups of unlisted ingredients are there?

33 How much less than 1 is the difference between $5\frac{2}{3}$ and $4\frac{7}{8}$?

34 How much more than 2 is the difference between $3\frac{4}{5}$ and $6\frac{3}{4}$?

35 Which is the greater and by how much:
the difference between $4\frac{7}{8}$ and $9\frac{1}{3}$ or the sum of $1\frac{3}{4}$ and $2\frac{3}{10}$?

ADDING AND SUBTRACTING DECIMALS

A dealer in animal foods has a stock of 21.8 tonnes of cattle food. That day he takes delivery of 10.75 tonnes and sells 15.5 tonnes. To find how much feed he has at the end of the day we must be able to subtract 15.5 from the sum of 21.8 and 10.75.

ADDITION OF DECIMALS

To add decimals we add in the usual way.

$$4.2 + 13.1 = 17.3$$

```
        tens  units  tenths
                4  .  2
    +    1      3  .  1
         ──────────────
         1      7  .  3
```
2 tenths + 1 tenth
= 3 tenths

$$5.3 + 6.8 = 12.1$$

```
                5  .  3
    +           6  .  8
         ──────────────
         1      2  .  1
         ı      ı
```
3 tenths + 8 tenths
= 11 tenths
= 1 unit and 1 tenth

The headings above the figures need not be written as long as we know what they are and the decimal points are in line (including the invisible point after a whole number, e.g. $4 = 4.0$).

EXERCISE 6J

Find **a** $32.6 + 1.7$ **b** $3 + 1.6 + 0.032 + 2.0066$

a

$$32.6 + 1.7 = 34.3$$

```
        32.6
    +    1.7
        ─────
        34.3
          ı
```

b

$$3 + 1.6 + 0.032 + 2.0066 = 6.6386$$

```
         3.0
         1.6
         0.032
    +    2.0066
        ───────
         6.6386
```

Find

1 $7.2 + 3.6$

2 $6.21 + 1.34$

3 $0.013 + 0.026$

4 $3.87 + 0.11$

5 $4.6 + 1.23$

6 $13.14 + 0.9$

7 $4 + 3.6$

8 $9.24 + 3$

9 $3.6 + 0.08$

10 $7.2 + 0.32 + 1.6$

11 $0.0043 + 0.263$

12 $0.002 + 2.1$

13 $0.000\,52 + 0.001\,24$

14 $0.068 + 0.003 + 0.06$

15 $4.62 + 0.078$

16 $0.32 + 0.032 + 0.0032$

17 $4.6 + 0.0005$

18 $16.8 + 3.9$

19 $1.62 + 2.078 + 3.1$

20 $7.34 + 6 + 14.034$

21 Add 0.68 to 1.7

22 Find the sum of 3.28 and 14.021

23 To 7.9 add 4 and 3.72

24 Evaluate $7.9 + 0.62 + 5$

25 Find the sum of 8.6, 5 and 3.21

Find the perimeter of the triangle (the perimeter is the distance all round).

1.6 cm   2.3 cm

2.6 cm

Perimeter $= 1.6 + 2.3 + 2.6$ cm
$\qquad = 6.5$ cm

$$
\begin{array}{r}
1.6 \\
2.3 \\
+\ 2.6 \\
\hline
6.5 \\
\hline
\end{array}
$$

26 Find the perimeter of the rectangle

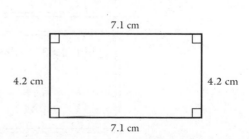

7.1 cm

4.2 cm   4.2 cm

7.1 cm

27 Find the total bill for three articles costing £5, £6.52 and £13.25.

28 Find the perimeter of the quadrilateral.

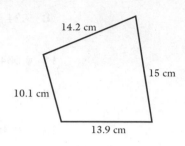

SUBTRACTION OF DECIMALS

Subtraction also may be done in the usual way, making sure that the decimal points are in line.

EXERCISE 6K

Find 24.2 − 13.7

$$\begin{array}{r} 24.2 \\ -\,13.7 \\ \hline 10.5 \end{array}$$

24.2 − 13.7 = 10.5

Find

1 6.8 − 4.3	**5** 0.0342 − 0.0021	**9** 102.6 − 31.2
2 9.6 − 1.8	**6** 17.23 − 0.36	**10** 7.32 − 0.67
3 32.7 − 14.2	**7** 3.273 − 1.032	**11** 54.07 − 12.62
4 0.62 − 0.21	**8** 0.262 − 0.071	**12** 7.063 − 0.124

It may be necessary to add noughts so that there are the same number of figures after the point in both cases.

Find 4.623 − 1.7

$$\begin{array}{r} 4.623 \\ -\,1.700 \\ \hline 2.923 \end{array}$$

4.623 − 1.7 = 2.923

Find 4.63 − 1.0342

$$\begin{array}{r} 4.6300 \\ -\,1.0342 \\ \hline 3.5958 \end{array}$$

4.63 − 1.0342 = 3.5958

Find $2 - 1.4$

$$
\begin{array}{r}
2.0 \\
- 1.4 \\
\hline
0.6
\end{array}
$$

$2 - 1.4 = 0.6$

Find

13 $3.26 - 0.2$

18 $7.98 - 0.098$

14 $3.2 - 0.26$

19 $7.098 - 0.98$

15 $14.23 - 11.11$

20 $3.2 - 0.428$

16 $6.8 - 4.14$

21 $11.2 - 0.0026$

17 $11 - 8.6$

22 $0.000\,32 - 0.000\,123$

23 $0.0073 - 0.0006$

28 $6 - 0.073$

24 $0.0073 - 0.006$

29 $7.3 - 0.06$

25 $0.006 - 0.000\,73$

30 $730 - 0.6$

26 $0.06 - 0.000\,73$

31 $0.73 - 0.000\,06$

27 $6 - 0.73$

32 $0.73 - 0.6$

33 Take 19.2 from 76.8

35 From 0.168 subtract 0.019

34 Subtract 1.9 from 10.2

36 Evaluate $7.62 - 0.81$

37 A piece of webbing is 7.6 m long.
If 2.3 m is cut off, how much is left?

38 The bill for two books came to £14.24. One book cost £3.72.
What was the cost of the other one?

39 Add 2.32 and 0.68 and subtract the result from 4.

40 To answer the question 'What length is left when I cut 2.5 cm and then 7 cm of tape from a roll 150 cm long?' Janet wrote

2.5 cm + 7 cm = 3.2 cm
150 cm − 3.2 cm = 147.8 cm

Her answer is wrong.

a Without checking her working, how do you know that her answer is obviously wrong?

b Find her mistakes.
Explain how you know they were mistakes.

EXERCISE 6L   Find the value of

1 8.62 + 1.7

2 8.62 − 1.7

3 3.8 − 0.82

4 0.08 + 0.32 + 6.2

5 5 − 0.6

6 100 + 0.28

7 100 − 0.28

8 0.26 + 0.026

9 0.26 − 0.026

10 78.42 − 0.8

11 38.2 + 1.68

12 38.2 − 1.68

13 0.84 + 2 + 200

14 16 + 1.6 + 0.16

15 1.4 − 0.81

16 0.02 − 0.013

17 0.062 + 0.32

18 6.83 − 0.19

19 17.2 + 20 + 1.62

20 9.2 + 13.21 − 14.6

21 The perimeter of the quadrilateral is 19 cm. What is the length of the fourth side?

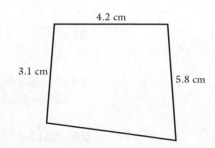

22 The bill for three meals was £9. The first meal cost £2.43 and the second £3.72.
How much was the cost of the third?

23 Sandra bought 3 magazines and a newspaper. The first magazine cost £2.75, the second cost 35 p more than the first, and the third magazine cost 39 p more than the second.
If she received 41 p change when she paid with a £10 note, how much did the newspaper cost?

24 A quadrilateral, like the one shown in question **21**, has a perimeter of 27 cm. The shortest side is 5.2 cm long; the next shortest side is 1.36 cm longer than this, and the length of the longest side is 50% more than the length of the shortest side.
How long is the fourth side?

MIXED EXERCISES

EXERCISE 6M

1 Add **a** $\frac{3}{5} + \frac{1}{5}$ **b** $\frac{1}{3} + \frac{1}{4} + \frac{7}{24}$

2 Find **a** $\frac{9}{16} - \frac{7}{16}$ **b** $\frac{7}{12} - \frac{1}{4}$ **c** $\frac{1}{2} + \frac{3}{7} - \frac{5}{8}$

3 Find $42 \div 9$ giving your answer as a mixed number.

4 Evaluate $1.5 + 6$

5 Find **a** $3\frac{1}{4} + 2\frac{2}{5}$ **b** $7\frac{3}{5} - 2\frac{7}{10}$

6 Find **a** $7.34 + 2.18$ **b** $0.54 + 0.049$

7 Find **a** $10 - 3.98$ **b** $4.2 - 0.972$

8 Subtract 6.02 from the sum of 1.27 and 18.59

9 In a class the weight of $\frac{1}{3}$ of the pupils is less than 50 kg and the weight of $\frac{5}{6}$ of the pupils is less than 60 kg. No pupil is exactly 50 kg.
What fraction of the pupils weigh between 50 kg and 60 kg?

EXERCISE 6N

1 Find **a** $\frac{1}{6} + \frac{5}{12}$ **b** $\frac{2}{3} + \frac{1}{5} + \frac{2}{15}$

2 Find **a** $\frac{7}{10} - \frac{1}{2}$ **b** $\frac{13}{21} - \frac{3}{7}$

3 Find the total bill for three books costing £4.26, £5 and £2.37.

4 Subtract 14.8 from 16.3.

5 The side of an equilateral triangle (all three sides are equal) is 14.4 cm. What is the length of the perimeter?

6 Change $5\frac{1}{6}$ into an improper fraction.

7 Find **a** $3\frac{3}{5} + 2\frac{2}{5}$ **b** $9\frac{1}{2} - 3\frac{2}{5}$

8 Find **a** $1.47 + 0.147$ **b** $3.96 + 11.2$

9 Find **a** $53.08 - 16.49$ **b** $0.937 - 0.088$

NUMBER PUZZLES

1

4.33	0.59	2.36	5.608	3.182	0.57	0.649
6.25	1.89	5.81	3.218	1.14	2.98	3.902
3.72	0.9	3.7	5.959	6.27	6.804	0.098
0.13	5.91	3.241	0.68	1.291	2.99	4.2

a Pair off as many numbers as possible so that all your number pairs add up to a number between 5 and 7. When your time is up your score is the sum of all the remaining numbers. The lower the score the better. (You will probably find it helpful to copy the list and cross out each pair of numbers that satisfies the condition.)

You have a time limit of two minutes for this exercise.

b Repeat the exercise to try to reduce your score.

c What is the lowest score possible?

d Try to write down twelve numbers written as three rows with four numbers in each row so that the rules given in part **a** apply and the lowest possible score is 0. If you think you've succeeded try it on a friend.

e Using the same list of decimals pair them off so that the difference between the pairs lies between 0 and 1.

2 If the sum of the numbers in all the rows, columns and diagonals of a square is the same the square is called a magic square.

For example in this magic square the total in every row, column and diagonal is 21.

9	8	4
2	7	12
10	6	5

a

		8.1
5.4	6.3	7.2
	10.8	

Fill in the blanks in this magic square if the total is always 18.9.

b

6.3	5.5	2.8
1.4	4.9	8.5
7.0	4.2	3.5

This magic square contains two wrong numbers. Find these wrong numbers and correct them.

c

15	10	30	60
40	50	16	90
14	11	20	70
10	80	13	12

Insert decimal points so that this is a magic square.

INVESTIGATION

In this investigation all fractions must be less than 1.
Look at the digits 2, 3, 4, 5, 6 and 7.

a Use two of these digits to make
   i the largest fraction possible   **ii** the smallest fraction possible.

b Use two of the digits on the top and another two on the bottom to make
   i the largest fraction possible   **ii** the smallest fraction possible.
   (Note: no digit can be used more than once in the same fraction.)

c If possible find two pairs of these digits to make two equivalent fractions.

d Make two fractions so that their difference is
   i the largest possible   **ii** the smallest possible.
   (Each digit can be used only once. The numerator and denominator must have the same number of digits.)

e Investigate what happens if you start with a different collection of six digits.

f What happens if the number of digits on the top and on the bottom can be different?

g What happens if you can have more than six digits to select from?

MORE ON DECIMALS

Discuss with a partner how the following problems can be solved.

A dealer has £350 to invest in radios costing £15.42 each. He wants to know how many he can afford to order.

Harri wants 36 lengths of 30 mm square timber, each 1.24 m long. The builders' merchant has a large quantity of this timber at a reduced price provided Harri is willing to take 6.1 m lengths. Harri wonders whether this is the cheapest way of buying what he wants. Would it be better to buy the exact lengths he needs at the normal price, which is 20% more per metre than the reduced price?

A chemist finds the easiest way to count tablets is by weighing them. He knows that 30 of a particular tablet weigh 22.5 g. He needs 100 tablets for a prescription.

EXERCISE 7A

1 Can you think of any occasion in the last week when you needed to multiply or divide decimals?

2 Which of the following people need to be able to multiply and divide decimals?

    a a carpenter   **c** a builder   **e** a banker

    b a shopkeeper   **d** an engineer   **f** a nurse?

Justify your answers.

3 In which of the following situations is it an advantage to be able to multiply and/or divide decimals?

    a Changing pounds into foreign money to go abroad

    b Getting the ingredients together to do some cooking

    c Buying tickets for a group of pupils to go to a concert

You should now be convinced that everyone needs to be able to multiply and divide decimals.

MULTIPLICATION
BY
10, 100, 1000, ...

Consider $32 \times 10 = 320$. Writing 32 and 320 in headed columns gives

hundreds	tens	units
	3	2
3	2	0

Multiplying by 10 has made the number of units become the number of tens, and the number of tens has become the number of hundreds, so that all the figures have moved one place to the left.

Consider 0.2×10. When multiplied by 10, tenths become units $\left(\frac{1}{10} \times 10 = 1 \right)$, so

units		tenths				units
0	.	2	$\times$	10	=	2

Again the figure has moved one place to the left.

Multiplying by 100 means multiplying by 10 and then by 10 again, so the figures move 2 places to the left.

	tens	units		tenths	hundredths	thousandths		
		0	.	4	2	6	$\times$	100
=	4	2	.	6				

Notice that the figures move to the left while the point stays put but without headings it looks as though the figures stay put and the point moves to the right.

When necessary we fill in an empty space with a nought.

units		tenths				hundreds	tens	units
4	.	2	$\times$ 100	=		4	2	0

EXERCISE 7B

1 Key in 54.2 on your calculator.

  a Multiply it by 10.

  b Multiply your answer by 10.

  c Multiply this answer by 10.

  What happens to the figures 542 each time you multiply by 10?

> Without using a calculator find the value of
>
> **a** 368 × 100   **b** 3.68 × 10   **c** 3.68 × 1000
>
> **a** 368 × 100 = 36 800
>
> **b** 3.68 × 10 = 36.8
>
> **c** 3.68 × 1000 = 3680

Find, without using a calculator

2 72 × 100   **5** 32.78 × 10   **8** 72.81 × 1000

3 8.24 × 10   **6** 0.043 × 100   **9** 0.0063 × 10

4 0.024 × 100   **7** 0.007 × 1000   **10** 3.74 × 100

11 A £1 coin is 2.81 mm thick. How high is a pile of 10 similar coins?

12 Hanwar is cutting pieces off a large ball of string.
Each piece is 14.5 cm long. He cuts off 100 pieces.
Find the total length of the string he cuts off.

**DIVISION BY
10, 100, 1000, …**

When we divide by 10, the hundreds become tens; the tens become units.

hundreds	tens	units				tens	units
6	4	0	÷	10	=	6	4

The figures move one place to the right and the number becomes smaller but it looks as though the decimal point moves to the left, e.g.

$$2.83 \div 10 = 0.283$$

To divide by 100 the point is moved two places to the left.
To divide by 1000 the point is moved three places to the left.

EXERCISE 7C

1 Key in 138.2 on your calculator.

  a Divide it by 10.   **c** Divide this answer by 10.

  b Divide your answer by 10.   **d** Divide your last answer by 10.

  What happens to the decimal point each time you divide by 10?

> Find, without using a calculator, the value of
>
> **a** 3.2 ÷ 10   **b** 43 ÷ 1000
>
> **a** 3.2 ÷ 10 = 0.32   ⟨ The decimal point moves 1 place to the left. ⟩
>
> **b** 43 ÷ 1000 = 0.043   ⟨ The decimal point moves 3 places to the left and we need to fill the 'empty' place with a nought. ⟩

Without using a calculator, find the value of

2 $266.8 \div 100$ **5** $27 \div 10$ **8** $426 \div 1000$

3 $76.37 \div 10$ **6** $5.8 \div 100$ **9** $16.8 \div 10$

4 $1.6 \div 100$ **7** $0.44 \div 10$ **10** $0.82 \div 100$

11 A pile of ten 20 p coins is 1.6 cm high. How thick is one 20 p coin?

12 When laid end to end 100 screws stretch a distance of 520 cm. What is the length of 1 screw?

The remaining questions are a mixture of multiplication and division.

Find

13 $1.6 \div 10$ **17** 14.3×100 **21** $140 \div 1000$

14 1.6×10 **18** 0.068×100 **22** 7.8×1000

15 $0.067 \div 10$ **19** $1.63 \div 100$ **23** $56 \div 100$

16 0.32×10 **20** 12.3×100 **24** 0.27×1000

25 $8.2 \div 100$ **27** $0.77 \div 100$ **29** $0.045 \div 1000$

26 $0.078 \div 100$ **28** $0.58 \div 1000$ **30** 0.041×1000

31 Sixty-seven metres of string is shared equally among 10 people. What length of string does each person get?

32 Find the total cost of 100 articles at £1.64 each.

33 Which of the following is the larger, and by how much:
0.47×100 or $560 \div 10$?

34 Which of the following is the smaller, and by how much:
$1.36 \div 10$ or 0.046×100?

35 Find the cost of 700 programmes at £1.25 each.

36 Multiply 2.7 by 100 and then divide the result by 1000.

37 Divide 740 by 100, and then divide the result by 10.

38 Add 14.9 to 3.45 and divide the result by 10.

39 Take 3.76 from 10 and multiply the result by 100.

40 A lorry is carrying 1000 boxes.
If each box weighs 1.34 kg what load is the lorry carrying?

41 100 cars, each 4.245 m long are parked nose to tail in a straight line.
How far is it from the front of the first car to the tail of the last one?

42 Which of the following is the larger, and by how much:
the difference between 4.6 and 15.9 multiplied by 10,
or the sum of 150.3 and 52 divided by 100?

43 A box contains 100 nails, each nail weighing 0.88 g.
Find the weight, in grams, of the nails in
a 1 box   **b** 100 boxes.

**MULTIPLICATION
BY WHOLE
NUMBERS**

A decimal can be multiplied by a whole number in the same way as one whole number can be multiplied by another. It is easier to keep the decimal point in the correct place when the calculation is set out in columns, e.g.

$$1.87 \times 6 = 11.22$$

$$
\begin{array}{r}
1.87 \\
\times 6 \\
\hline
11.22 \\
{\scriptstyle 5\ 4}
\end{array}
$$

EXERCISE 7D

Find

1 1.3×4 **5** 2.81×3 **9** 3.9×5

2 2.7×5 **6** 1.95×2 **10** 2.68×3

3 0.8×8 **7** 52.4×4 **11** 0.126×4

4 4.2×3 **8** 812.9×4 **12** 53.72×6

Find   **a** 8.56×300 **b** 0.63×700

a $8.56 \times 300 = 8.56 \times 3 \times 100$
$ = 25.68 \times 100$ First multiply by 3, then by 100.
$ = 2568$

b $0.63 \times 700 = 0.63 \times 7 \times 100$
$ = 4.41 \times 100$ First multiply by 7, then by 100.
$ = 441$

---

13 3.4×30 **16** 5.67×400 **19** 0.86×700

14 0.56×500 **17** 0.083×600 **20** 32.6×90

15 6.40×80 **18** 2.75×50 **21** 0.72×60

DIVISION BY
WHOLE
NUMBERS

We can see that   units   tenths   units   tenths

$$0 \,.\, 6 \;\div\; 2 \;=\; 0 \,.\, 3$$

because 6 tenths ÷ 2 = 3 tenths. So we may divide by a whole number in the usual way as long as we keep the figures in the correct columns and the points are in line.

EXERCISE 7E

Find the value of 6.8 ÷ 2

$$2 \,)\overline{6.8}$$
$$3.4$$

6.8 ÷ 2 = 3.4

Keep the figures and points in line.

Find the value of

1 0.4 ÷ 2   **5** 0.9 ÷ 9   **9** 42.6 ÷ 2

2 3.2 ÷ 2   **6** 0.95 ÷ 5   **10** 7.53 ÷ 3

3 0.63 ÷ 3   **7** 0.672 ÷ 3   **11** 6.56 ÷ 4

4 7.8 ÷ 3   **8** 26.6 ÷ 7   **12** 0.75 ÷ 5

13 A pile of four 1 p coins is 5.36 mm high.
    What is the thickness of one coin?

14 Eight boxed exercise benches are stacked one on top of the other
    and reach a height of 1.152 m. How deep is one box?

Sometimes it is necessary to fill in spaces with noughts.

Find   **a** 0.000 36 ÷ 3   **b** 0.45 ÷ 5   **c** 6.12 ÷ 3

a 0.000 36 ÷ 3 = 0.000 12

$$3 \,)\overline{0.000\,36}$$
$$0.000\,12$$

b 0.45 ÷ 5 = 0.09

$$5 \,)\overline{0.45}$$
$$0.09$$

c 6.12 ÷ 3 = 2.04

$$3 \,)\overline{6.12}$$
$$2.04$$

Find

15 0.057 ÷ 3   **18** 0.012 ÷ 6   **21** 1.232 ÷ 4

16 0.0065 ÷ 5   **19** 0.036 ÷ 6   **22** 0.6552 ÷ 6

17 0.008 72 ÷ 4   **20** 1.62 ÷ 2   **23** 0.0285 ÷ 5

24 $0.0076 \div 4$ **26** $0.5215 \div 5$ **28** $0.0636 \div 6$

25 $0.81 \div 9$ **27** $6.3 \div 7$ **29** $0.038 \div 2$

Find, without using a calculator, $194.6 \div 20$

$194.6 \div 10 = 19.46$

$19.46 \div 2 = 9.73$ (Divide by 10, then by 2.)

Find, without using a calculator

30 $26.4 \div 20$ **33** $105.6 \div 40$ **36** $19.5 \div 50$

31 $157.8 \div 30$ **34** $15.93 \div 90$ **37** $2.94 \div 70$

32 $40.2 \div 600$ **35** $1.944 \div 800$ **38** $59.1 \div 300$

It may be necessary to add noughts at the end of a number in order to finish the division.

Find   **a** $2 \div 5$ **b** $2.9 \div 8$

a $2 \div 5 = 0.4$ $5 \overline{)\,2.0\,}$

                        0.4

b $2.9 \div 8 = 0.3625$ $8 \overline{)\,2.9000\,}$

                            0.3625

Find the value of

39 $6 \div 5$ **43** $0.002 \div 5$ **47** $9.4 \div 4$

40 $7.4 \div 4$ **44** $7.1 \div 8$ **48** $0.5 \div 4$

41 $0.83 \div 2$ **45** $9.1 \div 2$ **49** $0.31 \div 8$

42 $3.6 \div 5$ **46** $0.0031 \div 2$ **50** $7.62 \div 4$

51 The perimeter (the distance all the way round the edge) of a square is 14.6 cm.
What is the length of a side?

52 Divide 32.6 into 8 equal parts.

53 Share 14.3 kg equally between 2 people.

54 The perimeter of a regular pentagon (a five-sided figure with all the sides equal) is 16 cm.
What is the length of one side?

DIVISION OF DECIMALS BY A TWO-DIGIT WHOLE NUMBER

Dividing a decimal by any whole number other than a single figure is best done using a calculator. Remember to estimate your answer as a check.

EXERCISE 7F

Use a calculator to find the value of

a $2.56 \div 16$ **b** $4.2 \div 24$

a Estimate: $3 \div 20 = 0.15$
∴ $2.56 \div 16 = 0.16$

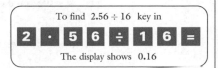

To find $2.56 \div 16$ key in

The display shows 0.16

b Estimate: $4 \div 20 = 0.2$
∴ $4.2 \div 24 = 0.175$

To find $4.2 \div 24$ key in

The display shows 0.175

First write down an estimate and then use a calculator to find the value of

1 $26.4 \div 24$ **4** $11.22 \div 22$ **7** $8.48 \div 16$

2 $2.1 \div 14$ **5** $90 \div 25$ **8** $5.2 \div 20$

3 $9.45 \div 21$ **6** $85.8 \div 26$ **9** $25.2 \div 36$

10 $34.54 \div 11$ **12** $54.4 \div 17$ **14** $23.4 \div 45$

11 $8.96 \div 28$ **13** $21.93 \div 51$ **15** $71.76 \div 23$

16 A carpet with an area of 9 square metres costs £310.50. How much is this a square metre?

17 Divide £18.09 among 27 children so that they each get the same amount.

18 A shopkeeper buys 32 compact discs for £159.68. How much each is this?

19 Sixteen screws weigh 19.56 g.
Find the weight of **a** 1 screw **b** 7 screws.

CHANGING FRACTIONS TO DECIMALS (EXACT VALUES)

Four friends went for a meal. Together with the tip, the total cost was £25. To find out how much each friend was due to pay we want to divide £25 by 4.

We can write this as £25 ÷ 4 or £$\frac{25}{4}$ but what does £$\frac{25}{4}$ mean? To find out how much each friend pays we work out £25 ÷ 4

$$\begin{array}{r} 6.25 \\ 4\overline{)\,25.00} \end{array}$$

i.e.   each friend pays £6.25, so £$\frac{25}{4}$ = £6.25,

i.e.   we have changed the fraction $\frac{25}{4}$ to the decimal 6.25.

We know that $\frac{3}{8}$ means 3 ÷ 8, and $\begin{array}{r}0.375\\8\overline{)\,3.000}\end{array}$...; i.e. $\frac{3}{8}$ = 0.375

> To change a fraction to a decimal,
> divide the bottom number into the top.

EXERCISE 7G

Express as a decimal   **a** $\frac{3}{4}$ **b** $1\frac{2}{25}$

a $\frac{3}{4} = 3 \div 4 = 0.75$ $\begin{array}{r}4\overline{)\,3.00}\\0.75\end{array}$

b $1\frac{2}{25} = 1 + 2 \div 25$ $\begin{array}{r}0.08\\25\overline{)\,2.00}\\200\end{array}$
$= 1 + 0.08$
$= 1.08$

Express the following fractions as decimals.

1 $\frac{1}{4}$ **3** $\frac{3}{5}$ **5** $\frac{1}{25}$ **7** $\frac{5}{8}$

2 $\frac{3}{8}$ **4** $\frac{5}{16}$ **6** $2\frac{4}{5}$ **8** $1\frac{7}{16}$

In questions **9** to **12** express each fraction as a decimal.

9 Everybody in the class sat a test and $\frac{7}{8}$ of the pupils passed.

10 A large crowd gathered to watch the opening of the new supermarket; $\frac{3}{4}$ of the crowd were women.

11 Hank inspected a delivery of apples. He found that $\frac{1}{16}$ of them were bruised.

12 During one cold night last winter $\frac{1}{8}$ of Kim's geraniums were killed by the frost.

STANDARD
DECIMALS AND
FRACTIONS

It is worth while knowing a few equivalent fractions, decimals and percentages.

For example

$$\frac{1}{2} = 0.5 = 50\% \qquad \frac{1}{4} = 0.25 = 25\% \qquad \frac{1}{8} = 0.125 = 12.5\%$$

Notice that $\qquad 0.4 = \frac{4}{10} = \frac{2}{5}$

EXERCISE 7H

Write the following decimals as fractions in their lowest terms, without any working if possible.

1 0.2	**4** 0.75	**7** 0.9	**10** 0.04
2 0.3	**5** 0.6	**8** 0.05	**11** 0.005
3 0.8	**6** 0.7	**9** 0.375	**12** 0.025

Write the following fractions as decimals.

13 $\frac{9}{10}$	**15** $\frac{4}{5}$	**17** $\frac{3}{100}$	**19** $\frac{5}{8}$
14 $\frac{1}{4}$	**16** $\frac{3}{8}$	**18** $\frac{3}{4}$	**20** $\frac{7}{100}$

MULTIPLICATION
BY A DECIMAL

So far we have multiplied a decimal by a multiple of 10 and by a whole number.
We have also divided a decimal by a multiple of 10 i.e. by 10, 100, 1000, ...

Using this knowledge enables us to multiply any two decimals together.

Suppose we wish to find 0.3×0.2

Since $0.2 = 2 \div 10$ $\qquad 0.3 \times 0.2 = 0.3 \times 2 \div 10$
$$= 0.6 \div 10$$
$$= 0.06$$

Similarly $\qquad 0.06 \times 0.08 = 0.06 \times 8 \div 100$
$$= 0.48 \div 100$$
$$= 0.0048$$

EXERCISE 7I

Find 0.05×0.7

$$0.05 \times 0.7 = 0.05 \times 7 \div 10 \qquad \boxed{0.7 = 7 \div 10}$$
$$= 0.35 \div 10$$
$$= 0.035$$

Find

1 0.04×0.2 **3** 0.003×6 **5** 0.001×0.3

2 0.1×0.1 **4** 3×0.02 **6** 0.4×0.001

7 Compare the number of decimal places in your answers with the number of decimal places in the numbers being multiplied. What do you notice?

In the examples above, if we add together the number of figures (including noughts) after the decimal points in the original two decimals, we get the number of figures after the point in the answer.

The number of figures after the point is called the number of decimal places. In the examples in Exercise 7I, 0.05 has two decimal places, 0.7 has one decimal place and the answer, 0.035, has three decimal places, which is the sum of two and one.

We can use this fact to work out 0.05×0.7 without writing 0.7 as $7 \div 10$. Multiply 5 by 7 ignoring the points; count up the number of decimal places after the points and then put the point in the correct position in the answer, writing in noughts where necessary, i.e. $0.05 \times 0.7 = 0.035$.
Any noughts that come after the point must be included when counting the decimal places.

EXERCISE 7J

Calculate the products   **a** 0.08×0.4 **b** 6×0.002

a   0.08 $\times$ 0.4 $=$ 0.032 $(8 \times 4 = 32)$
     (2 places)   (1 place)   (3 places)

b   6 $\times$ 0.002 $=$ 0.012 $(6 \times 2 = 12)$
     (0 places)   (3 places)   (3 places)

Calculate the following products. Do not use a calculator.

1 0.6×0.3 **4** 0.12×0.09 **7** 0.08×0.08

2 0.04×0.06 **5** 0.07×0.003 **8** 3×0.006

3 0.009×2 **6** 0.5×0.07 **9** 0.7×0.06

10 0.07×12 **12** 0.9×9 **14** 7×0.11

11 4×0.009 **13** 0.0008×11 **15** 0.04×7

16 Joe saws a length of wood into 7 equal pieces each 23.7 cm long. If each saw cut consumes 0.1 cm how long was the length of wood before he started sawing?

Noughts appearing in the multiplication, in the middle or at either end, must also be considered when counting the places.

Find   **a** 0.252×0.4 **b** 2.5×6 **c** 300×0.2

a $0.252 \times 0.4 = 0.1008$ 252
(3 places)   (1 place)   (4 places) $\times \underline{\quad 4}$
$\overline{1008}$

b $2.5 \times 6 = 15.0$ 25
(1 place)   (0 places)   (1 place) $\times \underline{\quad 6}$
$\overline{150}$

c $300 \times 0.02 = 6.00$ $300 \times 2 = 600$
(0 places)   (2 places)   (2 places)

Calculate the following products.

17 0.751×0.2 **20** 400×0.6 **23** 320×0.07

18 3.2×0.5 **21** 31.5×2 **24** 0.4×0.0055

19 0.35×4 **22** 5.6×0.02 **25** 0.5×0.006

26 1.6×0.4 **29** 4×1.6 **32** 0.16×4

27 1.6×0.5 **30** 5×0.016 **33** 0.0016×5

28 0.16×0.005 **31** $16\,000 \times 0.05$ **34** 310×0.04

35 A metal washer is 1.25 mm thick. Six washers are placed on each of the two supports of a gate in order to raise it.
By how much is the gate raised?

36 Find the cost of 10 articles at £43.50 each.

37 A sheet of paper is 0.065 mm thick.
How thick is a pile of 240 sheets?

38 Find the weight of 34 books if each book weighs 1.35 kg.

Calculate 0.26×1.3

$$0.26 \quad \times \quad 1.3 \quad = \quad 0.338$$
(2 places)   (1 place)   (3 places)

$$\begin{array}{r} 26 \\ \times\ 13 \\ \hline 78 \\ 260 \\ \hline 338 \end{array}$$

Calculate the following products.

39 4.2×1.6 **42** 310×1.4 **45** 0.082×0.034

40 52×0.24 **43** 1.68×0.27 **46** 0.0016×1600

41 0.68×0.14 **44** 13.2×2.5 **47** 0.34×0.31

48 A fire engine travels 0.675 miles in each minute.

How far will it travel in   **a** 4 minutes   **b** 12 minutes?

49 A taxi, that can travel at 0.6 miles per minute, had a call at 10.22 to pick up a fare at a passenger's home which was 4.5 miles away. The passenger wished to be taken to the station to catch a train due at 10.50. The taxi driver answered the call immediately, spent 1.5 minutes loading the passenger and her luggage, and drove the 6.5 miles to the station.

   a How long did the taxi take to get to the passenger?

   b How long did the taxi take to drive the passenger to the station?

   c At what time did the passenger arrive at the station?

   d If 3.25 minutes were allowed to get from the taxi to the station platform, did the passenger catch the train?
   If so, how much time did she have to spare?

**CORRECTING TO
A GIVEN NUMBER
OF DECIMAL
PLACES**

Often we need to know only the first few figures of a decimal. For instance, if we measure a length with an ordinary ruler we usually need an answer to the nearest $\frac{1}{10}$ cm and are not interested, or cannot see, how many $\frac{1}{100}$ cm are involved.

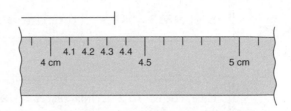

Look at this enlarged view of the end of a line which is being measured. We can see that with a more accurate measure we might be able to give the length as 4.34 cm. However on the given ruler we would probably measure it as 4.3 cm because we can see that the end of the line is nearer 4.3 than 4.4. We cannot give the exact length of the line but we can say that it is 4.3 cm long to the nearest $\frac{1}{10}$ cm. We write this as 4.3 cm correct to 1 decimal place.

Consider the numbers 0.62, 0.622, 0.625, 0.627 and 0.63. To compare them we write 0.62 as 0.620 and 0.63 as 0.630 so that each number has 3 figures after the point i.e. 3 decimal places.

$$0.620$$
$$0.622$$
$$0.625$$
$$0.627$$
$$0.630$$

When we write them in order in a column we can see that 0.622 is nearer to 0.620 than to 0.630 while 0.627 is nearer to 0.630 so we write

$$0.622 = 0.62 \qquad (\text{correct to 2 decimal places})$$

$$0.627 = 0.63 \qquad (\text{correct to 2 decimal places})$$

It is not so obvious what to do with 0.625 as it is halfway between 0.62 and 0.63. To save arguments, if the figure after the cut-off line is 5 or more we add 1 to the figure before the cut-off line, i.e. we round the number *up*, so we write

$$0.625 = 0.63 \qquad (\text{correct to 2 decimal places})$$

EXERCISE 7K

> Give 10.9315 correct to
>
> **a** the nearest whole number   **b** 1 decimal place
>
> **a** 10.9315 = 11   (correct to the nearest whole number)
>
> **b** 10.9315 = 10.9   (correct to 1 decimal place)

> Give, correct to 2 decimal places   **a** 4.699   **b** 0.007
>
> **a** 4.699 = 4.70   (correct to 2 decimal places)
>
> > Note the zero in the second decimal place shows
> > that the number *has* been corrected to 2 dp.
>
> **b** 0.007 = 0.01   (correct to 2 decimal places)

Give the following numbers correct to 2 decimal places.

1 0.328   **6** 0.6947   **11** 0.178

2 0.322   **7** 0.8351   **12** 1.582

3 1.2671   **8** 3.927   **13** 4.995

4 2.345   **9** 0.0084   **14** 0.0115

5 0.0416   **10** 3.9999   **15** 8.0293

Give the following numbers correct to the nearest whole number.

16 13.9   **21** 6.783   **26** 58.4

17 6.34   **22** 109.7   **27** 0.98

18 26.5   **23** 6.145   **28** 15.29

19 2.78   **24** 74.09   **29** 152.9

20 4.45   **25** 3.9999   **30** 26.49

Give the following numbers correct to the number of decimal places indicated in the brackets.

31 1.784 (1)

36 1.639 (2)

32 42.64 (1)

37 1.639 (1)

33 1.0092 (2)

38 1.639 (nearest whole number)

34 0.0942 (2)

39 343.4984 (2)

35 0.7345 (1)

40 3.4984 (1)

If we are asked to give an answer correct to a certain number of decimal places, we work out one more decimal place than is asked for. Then we can find the size of the last figure required.

EXERCISE 7L

Find $4.28 \div 6$ giving your answer correct to 2 decimal places.

$$6 \,)\, \overline{4.280 \ldots}$$
$$0.713 \ldots$$

We need to give the answer to 2 decimal places so we carry on dividing until we know the third decimal place.

$4.28 \div 6 = 0.713 \ldots$
$\qquad = 0.71$ (correct to 2 decimal places)

Calculate $302 \div 14$ correct to 1 decimal place.

We can stop the division when we know the 2nd decimal place.

$$\begin{array}{r} 21.57 \ldots \\ 14\overline{)302.00} \\ 28 \\ \overline{22} \\ 14 \\ \overline{80} \\ 70 \\ \overline{100} \\ 98 \end{array}$$

$302 \div 14 = 21.57 \ldots$
$\qquad = 21.6$ (correct to 1 decimal place)

Calculate, giving your answers correct to 2 decimal places.

1 $0.496 \div 3$

4 $25.68 \div 9$

7 $5.68 \div 24$

2 $6.49 \div 7$

5 $2.35 \div 15$

8 $3.85 \div 11$

3 $12.2 \div 6$

6 $0.99 \div 21$

9 $1.83 \div 8$

So far, for much of the time we have used pencil and paper methods. However, it is much easier to use a calculator, provided you make a rough check to see that your answers are sensible.

Use a calculator to find

a $32.9 \div 21$ correct to 2 decimal places

b $17.89 \div 37$ correct to 1 decimal place.

a Estimate:
$30 \div 20 = 3 \div 2 = 1.5$

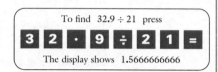

To find $32.9 \div 21$ press

The display shows 1.5666666666

$32.9 \div 21 = 1.57$ (correct to 2 decimal places)

b Estimate:
$20 \div 40 = 2 \div 4 = 0.5$

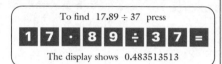

To find $17.89 \div 37$ press

The display shows 0.483513513

$17.89 \div 37 = 0.5$ (correct to 1 decimal place)

Use a calculator to find each value correct to the number of decimal places indicated in the brackets.

10 $32.9 \div 8$ (1)   **14** $124 \div 17$ (1)   **18** $213 \div 22$ (2)

11 $402 \div 7$ (1)   **15** $0.45 \div 12$ (2)   **19** $519 \div 19$ (2)

12 $9.76 \div 11$ (1)   **16** $15.1 \div 16$ (1)   **20** $0.321 \div 17$ (2)

13 $2.3 \div 11$ (2)   **17** $2.584 \div 16$ (2)   **21** $54 \div 23$ (1)

22 Seven brothers and sisters are given £100 to be divided equally among them.
Correct to the nearest penny, how much will each one receive?
If each brother and sister received this amount would £100 be enough?
Give reasons for your answer.

23 Express each of the fractions $\frac{4}{5}$, $\frac{6}{7}$, $\frac{13}{15}$ and $\frac{15}{19}$ as a decimal correct to 2 decimal places.
Hence write these fractions in order of size with the smallest first.

24 Express $\frac{3}{7}$, $\frac{5}{12}$ and $\frac{6}{13}$ as decimals correct to 2 decimal places. Hence write them in order of size with the smallest first.

25 Write the numbers 0.82, $\frac{9}{11}$, 0.801 and $\frac{19}{23}$ as decimals, correct to 3 decimal places where necessary, and write them in order of size with the largest first.

26 Look at the numbers 0.7, $\frac{7}{11}$, $\frac{14}{19}$ and 0.8. Decide, without using a calculator,

a which is the largest number **b** which is the smallest number. Can you give reasons for your answers?

27 Four pupils were asked to put the numbers $\frac{5}{9}$, $\frac{3}{5}$, 0.45 and $\frac{11}{17}$ in order of size, smallest first. Without using a calculator decide which of the following answers is probably correct.

a Sheena's 0.45, $\frac{5}{9}$, $\frac{3}{5}$, $\frac{11}{17}$ **c** Diane's 0.45, $\frac{11}{17}$, $\frac{3}{5}$, $\frac{5}{9}$

b Eddy's $\frac{3}{5}$, 0.45, $\frac{5}{9}$, $\frac{11}{17}$ **d** Jim's 0.45, $\frac{5}{9}$, $\frac{11}{17}$, $\frac{3}{5}$

28 Estimates suggest that the average man uses 12.3 calories per minute when swimming. How many calories, correct to 1 decimal place, are used when a man swims for

a 10 minutes **b** $22\frac{1}{2}$ minutes **c** $13\frac{1}{4}$ minutes?

DIVISION BY DECIMALS

$0.012 \div 0.06$ can be written as $\frac{0.012}{0.06}$. We know how to divide by a whole number so we need to find an equivalent fraction with denominator 6 instead of 0.06. Now $0.06 \times 100 = 6$. Therefore we multiply the numerator and denominator by 100.

$$\frac{0.012}{0.06} = \frac{0.012 \times 100}{0.06 \times 100}$$
$$= \frac{1.2}{6}$$
$$= 0.2$$

To divide by a decimal, the denominator must be made into a whole number but the numerator need not be. We can write, for short,

$$0.012 \div 0.06 = \frac{0.012}{0.06} = \frac{1.2}{6} \quad (\text{keeping the points in line})$$

the dotted line indicating where we want the point to be so as to make the denominator a whole number.

EXERCISE 7M

Find the exact value of   **a** $0.024 \div 0.6$ **b** $64 \div 0.08$

a $0.024 \div 0.6 = \dfrac{0.024}{0.6}$

$ = \dfrac{0.24}{6}$

$ = 0.04$

$$6\,)\,\overline{0.24}$$
$$0.04$$

b $64 \div 0.08 = \dfrac{64.00}{0.08}$

$ = \dfrac{6400}{8}$

$ = 800$

$$8\,)\,\overline{6400}$$
$$800$$

Find the exact answers to the following questions.

1 $0.04 \div 0.2$ **4** $90 \div 0.02$ **7** $3.6 \div 0.06$

2 $4 \div 0.5$ **5** $0.48 \div 0.04$ **8** $3 \div 0.6$

3 $0.8 \div 0.04$ **6** $0.032 \div 0.2$ **9** $6.5 \div 0.5$

10 $72 \div 0.9$ **13** $0.9 \div 0.009$ **16** $0.000\,068\,4 \div 0.04$

11 $1.08 \div 0.003$ **14** $0.92 \div 0.4$ **17** $20.8 \div 0.0004$

12 $16.8 \div 0.8$ **15** $0.001\,32 \div 0.11$ **18** $4.8 \div 0.08$

19 How many plastic washers, each 0.4 mm thick are needed to separate two pieces of metal by 1.6 mm ?

20 The height of one fence post is 1.8 m. The total of the heights of a bundle of fence posts is 39.6 m.
How many posts are there in the bundle ?

21 Eddie Davis keeps careful rainfall records. During one period last winter the total rainfall was 29.44 cm which, if the same quantity fell each day, would be 0.32 cm per day.
How many days did this period cover ?

Use a calculator to find the exact value of $0.256 \div 1.6$

Press   **· 2 5 6 ÷ 1 · 6 =**

The display shows 0.16

$0.256 \div 1.6 = 0.16$

Use a calculator to find the exact value of

22 $1.76 \div 2.2$ **25** $34.3 \div 1.4$ **28** $9.8 \div 1.4$

23 $144 \div 0.16$ **26** $10.24 \div 3.2$ **29** $0.168 \div 0.14$

24 $0.496 \div 1.6$ **27** $0.0204 \div 0.017$ **30** $0.192 \div 2.4$

31 How many coins each 0.73 mm thick are there in a pile 8.76 mm high?

32 The product of two numbers is 9.24. One of the numbers is 3.85. What is the other?

33 A company pays a dividend of 2.55 pence per share. Roy Hately receives £14.28.
How many shares does Roy have?

34 How many screws each weighing 1.26 g have a total weight of 79.38 g?

35 The thickness of a 350 page book (i.e. the book has 175 sheets or leaves) is 12.25 mm.

     a Find the thickness of
         i 1 sheet   **ii** 45 sheets   **iii** a ream (500 sheets).
     b How thick is a book with
         i 120 pages   **ii** 440 pages?
     c How many sheets have a total thickness of
         i 3.5 mm   **ii** 26.25 mm?
     d How many pages are there in a book whose thickness is
         i 12.6 mm   **ii** 8.96 mm?

EXERCISE 7N

Without using a calculator, find the value of $16.9 \div 0.3$ giving your answer correct to 1 decimal place.

$$16.9 \div 0.3 = \frac{16.9}{0.3} = \frac{169}{3} \qquad\qquad 3\overline{)169.00}$$
$$56.33\ldots$$

$$= 56.3|3 \ldots$$
$$= 56.3 \quad (\text{correct to 1 decimal place})$$

Without using a calculator, find each value correct to 2 decimal places.

1 $3.8 \div 0.6$ **4** $1.25 \div 0.03$

2 $0.59 \div 0.07$ **5** $0.0024 \div 0.09$

3 $15 \div 0.9$ **6** $0.65 \div 0.7$

7 $0.123 \div 6$ **10** $0.23 \div 0.007$

8 $2.3 \div 0.8$ **11** $16.2 \div 0.8$

9 $90 \div 11$ **12** $85 \div 0.3$

Use a calculator to give, correct to 2 decimal places,

a $0.52 \div 0.21$ **b** $56.79 \div 1.6$

a Estimate: $0.52 \div 0.21 \approx 0.5 \div 0.2 = 5 \div 2 \approx 3$

> When we estimate, all we want is a rough idea of the size of the answer, so we can also round the results of calculations when finding an estimate. This is why we give $5 \div 2 \simeq 3$.

Press   **0** **.** **5** **2** **÷** **0** **.** **2** **1** **=**

The display shows $2.47619\ldots$

$\therefore \quad 0.52 \div 0.21 = 2.48 \quad (\text{correct to 2 decimal places})$

b Estimate $56.79 \div 1.6 \approx 60 \div 2 = 30$

Press   **5** **6** **.** **7** **9** **÷** **1** **.** **6** **=**

The display shows 35.49375

$\therefore \quad 56.79 \div 1.6 = 35.49 \quad (\text{correct to 2 decimal places})$

First estimate and then use a calculator to find the value of each division. Give each answer correct to **a** 1 decimal place **b** 2 decimal places.

13 $47.3 \div 18$

14 $5.33 \div 1.08$

15 $0.653 \div 1.04$

16 $0.041 \div 0.13$

17 $3.65 \div 0.92$

18 $64.07 \div 12.44$

19 $0.0734 \div 0.119$

20 $17.83 \div 73.65$

MIXED MULTIPLICATION AND DIVISION

EXERCISE 7P

Without using a calculator and giving your answers exactly, calculate

1 0.48×0.3

2 $0.48 \div 0.3$

3 2.56×0.02

4 $2.56 \div 0.02$

5 $3.6 \div 0.8$

6 9.6×0.6

7 0.0042×0.03

8 $0.0042 \div 0.03$

9 16.8×0.4

10 $1.68 \div 0.4$

11 20.4×0.6

12 5.04×0.06

Use a calculator to find, correct to 2 decimal places

13 4.87×9.04

14 $20.08 \div 8.85$

15 $67.9 \div 126.4$

16 $15.64 \div 8.4$

17 $0.074 \div 0.22$

18 0.555×3.27

19 0.176×1.88

20 31.45×0.077

21 $9.004 \div 0.435$

22 Which is the larger answer

    a 3.6×6 or $36 \div 6$

    b 0.36×6 or $3.6 \div 0.6$

    c 0.9×0.7 or 0.8×0.8?

23 Fill in the blanks in the following calculations.

    a $5.2 \times \square = 19.24$

    b $17.38 \div \square = 4.4$

    c $0.294 \times 0.06 = \square$

    d $\square \div 0.35 = 1.44$

    e $0.93 \times 2.72 - \square = 0$

    f $\square \times 7.32 = 4.758$

**MIXED
EXERCISES**

Do not use a calculator for these exercises except when you are asked to do so.

EXERCISE 7Q

1 Divide 0.082 by 4.

2 Multiply 0.0301 by 100.

3 Express $\frac{7}{8}$ as a decimal.

4 Divide 1.5 by 25.

5 Give the following numbers correct to 2 decimal places.

  a 3.126   **b** 0.075   **c** 2.999

6 Multiply 3.2 by 1.4.

7 Which is the larger, 6.6 or $6\frac{2}{3}$? Why?

8 Use a calculator to find

  a 50.6 ÷ 2.2   **b** 0.683 × 1.8

9 Last season Jake took 35 wickets for 327 runs.
How many runs was this for each wicket? Give your answer correct to 2 decimal places.

EXERCISE 7R

1 Find **a** 42.64 × 10   **b** 37.43 ÷ 100

2 Find **a** 6.74 × 300   **b** 0.81 ÷ 9

3 Use a calculator to find   **a** 40.8 ÷ 23   **b** 1.62 ÷ 36

4 Find   **a** 0.008 × 6   **b** 0.5 × 0.12   **c** 230 × 1.4

5 Give 17.906 correct to
  a the nearest whole number   **b** 1 decimal place   **c** 2 decimal places

6 Find   **a** 5.92 ÷ 7   **b** 16.4 ÷ 9,
giving each answer correct to 2 decimal places.

7 Find the exact value of   **a** 2.56 ÷ 3.2   **b** 0.126 ÷ 1.8

8 Use a calculator to decide which is the larger, and by how much, 0.52 × 0.6 or 0.55 × 0.55.

9 Sheets of metal, 0.125 cm thick, are piled one on top of the other until they reach a height of 36.5 cm.
Use a calculator to find the number of sheets in the pile.

Recurring decimals

If we divide 2 by 5 we get an exact answer, namely 0.4
However, if we divide 2 by 3 the answer is 0.666 666 6 ... for ever,
i.e. $2 \div 3$ does not give an exact decimal. We say that the 6 recurs.
Some of the numbers we get when we divide one whole number by
another are interesting and are worth investigating.
(You should not use a calculator for this work before you have used
   pencil and paper methods.)

a Find $0.4 \div 7$. Continue working until you have at least 12 non-
zero figures in your answer. Use a calculator to check the first 8
non-zero figures of your answer.
What do you notice about the pattern of figures in your answer?

b Now try $0.2 \div 7$. Do you get the same figures in the same order?
How does this answer differ from the answer you got in part **a**?

c Can you find any other decimals which, when divided by 7, give
the same pattern of figures, but start in a different place in the
pattern?

d Now try any numbers larger than 1 that 7 does not divide into
exactly.
For example $1.5 \div 7$ and $2.2 \div 7$.

e Is it true to say that every decimal, when divided by 7, gives either
an exact answer, or a recurring answer that involves the same cycle
of figures?

f Repeat parts **a** to **e** but divide by 9 instead of 7.

g What happens if you divide by the other odd prime numbers less
than 10?

h What happens if you divide by the even numbers less than 10?

METRIC UNITS

When I want to order a new blind for a window, I need to give the width and height of the window. I can give these measurements by saying

- as wide as my desk and from halfway up the wall to the top of the door
- 96 centimetres wide and 102 centimetres high.

The first measurements are almost useless because the manufacturers do not know the width of my desk or the height of my wall or door. If they are accepted, the blind is unlikely to fit the window!

The second measurements are understandable and can be used to make a blind that is the right size. This is because a centimetre is a standard unit of length, i.e. other people know how long a centimetre is.

Whenever we want to communicate measurements to other people, it is sensible to use standard measures. We will then be understood.

UNITS OF LENGTH

The basic unit of length is the metre (m).
(A standard bed is about 2 m long.)

A metre is not a useful unit for measuring either very large things or very small things so we need larger units and smaller units.

We get the larger unit by multiplying the metre by 1000.

> 1000 metres is called 1 kilometre (km)

(It takes about 15 minutes to walk a distance of 1 km.)

We get smaller units by dividing the metre into 100 parts or into 1000 parts.

> $\frac{1}{100}$ of a metre is called 1 centimetre (cm)

> $\frac{1}{1000}$ of a metre is called 1 millimetre (mm)

(You can see centimetres and millimetres on your ruler.)

EXERCISE 8A

1 Discuss which metric unit you would use to measure

   a the length of your classroom

   b the length of your pencil

   c the length of a soccer pitch

   d the distance from Manchester to London

   e the length of a page in this book

   f the thickness of your exercise book.

2 You need a new table cloth. You see a cloth that you like. The label says that it measures 1246 mm by 2560 mm. You know that your table measures about 1 metre by 2 metres.
Discuss what you need to know so that you can judge if this cloth will cover your table.

ESTIMATING
LENGTHS

If you are asked how wide a particular book is, can you give a quick estimate of its width without having to measure it?
If you are told that a poster is 588 mm long, does this mean anything to you?
The skills that enable you to answer 'yes' to these questions come with experience of measuring and estimating.

EXERCISE 8B

1 Use your ruler to draw a line of length

a 10 cm	**e** 20 mm	**i** 25 mm
b 3 cm	**f** 40 cm	**j** 16 mm
c 15 cm	**g** 15 mm	**k** 5 cm
d 50 mm	**h** 12 cm	**l** 75 mm

2 Estimate the length, in centimetres, of the following lines.

   a ————————————————

   b ——————————————————————

   c ———————

   d —————————————————————

   e ——————————————————————————

Now use your ruler to measure each line.

3 Estimate the length, in millimetres, of the following lines.

a ———— **d** ——————

b —— **e** ——

c —

Now use your ruler to measure each line.

4 Use a straight edge (not a ruler with a scale) to draw a line that is approximately

a 10 cm long **b** 5 cm long **c** 15 cm long **d** 20 cm long

Now measure each line to see how good your approximation was.

5 Estimate the length and width of your exercise book in centimetres. Check your estimates by measuring the length and width of your exercise book. Draw a rough sketch of your book with the measurements on it.
Find the perimeter (the distance all round the edge) of your book.

6 Each side of a square is 10 cm long.
Draw a rough sketch of the square with the measurements on it.
Calculate the perimeter of the square.

7 A sheet is 200 cm wide and 250 cm long.

a What is the perimeter of the sheet?

b Will this sheet cover your bed?

c Will it cover the table you last sat at?

Explain your answers to parts **b** and **c**.

Questions **8** to **10** are group activities.

8 Estimate the width of your classroom in metres.
Check the estimate by measuring.

9 Estimate the length of your classroom in metres.
Now measure it to check your estimate.

10 a Measure the height of the door into your classroom. Use this to estimate the height of the room.

b Discuss how you can use the known height of a person to estimate the heights of some everyday objects, e.g. a double decker bus.

CHANGING FROM
LARGE UNITS TO
SMALLER UNITS

Which is longer: a tile 25 cm long or a tile 246 mm long?
It is easier to compare lengths if they are both measured in the same unit.

The metric units of length are the kilometre, the metre, the centimetre and the millimetre.

The relationships between them are

1 km = 1000 m	1 m = 100 cm
1 m = 1000 mm	1 cm = 10 mm

EXERCISE 8C

Express 3 km in metres.

$3 \, \text{km} = 3 \times 1000 \, \text{m}$
$\quad\quad\;\; = 3000 \, \text{m}$

Express 3.5 m in centimetres.

$3.5 \, \text{m} = 3.5 \times 100 \, \text{cm}$
$\quad\quad\;\; = 350 \, \text{cm}$

Express the given quantity in terms of the unit in brackets.

1 2 m (cm)   **5** 12 km (m)   **9** 3 m (mm)

2 5 km (m)   **6** 15 cm (mm)   **10** 2 km (mm)

3 3 cm (mm)   **7** 6 m (mm)   **11** 5 m (cm)

4 4 m (cm)   **8** 1 km (cm)   **12** 7 m (mm)

13 1.5 m (cm)   **17** 1.9 m (mm)   **21** 3.8 cm (mm)

14 2.3 cm (mm)   **18** 3.5 km (m)   **22** 9.2 m (mm)

15 4.6 km (m)   **19** 2.7 m (cm)   **23** 2.3 km (m)

16 3.7 m (mm)   **20** 1.9 km (cm)   **24** 8.4 m (cm)

25 Which is longer, a piece of wood 24 cm long or a piece 236 mm long?

26 Which piece of tape is longer – one that is 3.5 m long or one that is 1335 mm long?

27 Arrange the following lengths in order of size with the shortest first.

$$25\,m \qquad 156\,cm \qquad 2889\,mm \qquad 3.8\,m \qquad 0.57\,m$$

UNITS OF MASS

The mass of an object is the quantity of matter in it; the weight of an object is the pull of the earth on it and this is what we feel. When we put a bag of potatoes on a set of scales we are measuring its weight but the units on the scales are units of mass. If the reading is 2.5 kg, the mass is 2.5 kg but, in everyday language, we say that the bag of potatoes weighs 2.5 kg, or that the weight is 2.5 kg.

The most familiar units used for weighing are the kilogram (kg) and the gram (g).

Most groceries that are sold in tins or packets have weights given in grams. For example the weight of the most common packet of butter is 250 g and one eating apple weighs roughly 100 g. So we see that the gram is a small unit of weight. Kilograms are used to give the weights of sugar and flour; the weight of the most common bag of sugar is 1 kg and the most common bag of flour weighs 1.5 kg.

For weighing large loads (coal or steel, for example) a larger unit of weight is needed, and we use the tonne (t). For weighing very small quantities (for example the weight of a particular drug in one pill) we use the milligram (mg).

The relationships between these weights are

$$1\,t = 1000\,kg$$
$$1\,kg = 1000\,g$$
$$1\,g = 1000\,mg$$

EXERCISE 8D

Express 2.35 t in grams.

$$
\begin{aligned}
2.35\,t &= 2.35 \times 1000\,kg \\
&= 2350\,kg \\
&= 2350 \times 1000\,g \\
&= 2\,350\,000\,g
\end{aligned}
$$

First change tonnes to kilograms and then change kilograms to grams.

Express each quantity in terms of the unit given in brackets.

1	12 t (kg)	**5**	1 kg (mg)	**9**	4 kg (g)
2	3 kg (g)	**6**	13 kg (g)	**10**	2 kg (mg)
3	5 g (mg)	**7**	6 g (mg)	**11**	3 t (kg)
4	1 t (g)	**8**	2 t (g)	**12**	4 g (mg)
13	1.5 kg (g)	**17**	5.2 kg (mg)	**21**	7.3 g (mg)
14	2.7 t (kg)	**18**	0.6 g (mg)	**22**	0.3 kg (mg)
15	1.8 g (mg)	**19**	11.3 t (kg)	**23**	0.5 t (kg)
16	0.7 t (kg)	**20**	2.5 kg (g)	**24**	0.8 g (mg)

CHANGING FROM SMALL UNITS TO BIGGER UNITS

EXERCISE 8E

Express 400 cm in metres.

$$400 \, \text{cm} = 400 \div 100 \, \text{m}$$
$$= 4 \, \text{m}$$

Express 580 g in kilograms.

$$580 \, \text{g} = 580 \div 1000 \, \text{kg}$$
$$= 0.58 \, \text{kg}$$

In questions **1** to **20**, express the first quantity in terms of the unit given in brackets.

1	300 mm (cm)	**6**	72 m (km)
2	6000 m (km)	**7**	12 cm (m)
3	150 cm (m)	**8**	88 mm (cm)
4	250 mm (cm)	**9**	1250 mm (m)
5	1600 m (km)	**10**	2850 m (km)

11 1500 kg (t)   **<u>16</u>** 86 kg (t)

12 3680 g (kg)   **<u>17</u>** 560 g (kg)

13 1500 mg (g)   **<u>18</u>** 28 mg (g)

14 5020 g (kg)   **<u>19</u>** 190 kg (t)

15 3800 kg (t)   **<u>20</u>** 86 g (kg)

21 Which is heavier, a locomotive weighing 86 tonnes or a loaded lorry weighing 39 800 kg?

22 Put these weights in order of size with the lightest first.

3 t   900 kg   279 000 kg   279 000 g

23 Put these weights in order with the heaviest first.

7900 mg   2.5 g   0.25 kg   79 g

ADDING AND
SUBTRACTING
METRIC
QUANTITIES
Jane has several items to carry home. To find out the total weight of these items, she must add their weights.

Before we can add or subtract *any* two quantities, they must both be expressed in the same unit.

Now 250 g = 0.25 kg and 454 g = 0.454 kg,
so Jane has to carry (0.25 + 1.5 + 0.454) kg, i.e. 2.204 kg

EXERCISE 8F

Find $1\,\text{kg} + 158\,\text{g}$ in   **a** grams   **b** kilograms.

a $1\,\text{kg} = 1000\,\text{g}$
$\therefore \quad 1\,\text{kg} + 158\,\text{g} = 1158\,\text{g}$

b $158\,\text{g} = 158 \div 1000\,\text{kg}$
$= 0.158\,\text{kg}$
$\therefore \quad 1\,\text{kg} + 158\,\text{g} = 1.158\,\text{kg}$

Find the sum of $5\,\text{m}$, $4\,\text{cm}$ and $97\,\text{mm}$ in
a metres   **b** centimetres.

a $4\,\text{cm} = 4 \div 100\,\text{m} = 0.04\,\text{m}$
$97\,\text{mm} = 97 \div 1000\,\text{m} = 0.097\,\text{m}$
$\therefore \quad 5\,\text{m} + 4\,\text{cm} + 97\,\text{mm} = (5 + 0.04 + 0.097)\,\text{m}$
$= 5.137\,\text{m}$

b $5\,\text{m} = 5 \times 100\,\text{cm} = 500\,\text{cm}$
$97\,\text{mm} = 97 \div 10\,\text{cm} = 9.7\,\text{cm}$
$\therefore \quad 5\,\text{m} + 4\,\text{cm} + 97\,\text{mm} = (500 + 4 + 9.7)\,\text{cm}$
$= 513.7\,\text{cm}$

Find, giving your answer in metres

1 $5\,\text{m} + 86\,\text{cm}$ **4** $51\,\text{m} + 3\,\text{km}$

2 $92\,\text{cm} + 115\,\text{mm}$ **5** $36\,\text{cm} + 87\,\text{mm} + 520\,\text{cm}$

3 $3\,\text{km} + 136\,\text{cm}$ **6** $120\,\text{mm} + 53\,\text{cm} + 4\,\text{m}$

Find, giving your answer in millimetres

7 $36\,\text{cm} + 80\,\text{mm}$ **10** $2\,\text{m} + 45\,\text{cm} + 6\,\text{mm}$

8 $5\,\text{cm} + 5\,\text{mm}$ **11** $3\,\text{cm} + 5\,\text{m} + 2.9\,\text{cm}$

9 $1\,\text{m} + 82\,\text{cm}$ **12** $34\,\text{cm} + 18\,\text{mm} + 1\,\text{m}$

Find, giving your answer in grams

13 $3\,\text{kg} + 250\,\text{g}$ **16** $1\,\text{kg} + 0.9\,\text{kg} + 750\,\text{g}$

14 $5\,\text{kg} + 115\,\text{g}$ **17** $116\,\text{g} + 0.93\,\text{kg} + 680\,\text{mg}$

15 $5.8\,\text{kg} + 9.3\,\text{kg}$ **18** $248\,\text{g} + 0.06\,\text{kg} + 730\,\text{mg}$

Find, expressing your answer in kilograms,

19 $2\,t + 580\,kg$

20 $1.8\,t + 562\,kg$

21 $390\,g + 1.83\,kg$

22 $1.6\,t + 3.9\,kg + 2500\,g$

23 $1.03\,t + 9.6\,kg + 0.05\,t$

24 $5.4\,t + 272\,kg + 0.3\,t$

Find, expressing your answer in the unit given in brackets

25 $8\,m - 52\,cm\,(\,cm\,)$

26 $52\,mm + 87\,cm\,(\,m\,)$

27 $1.3\,kg - 150\,g\,(\,g\,)$

28 $1.3\,m - 564\,mm\,(\,cm\,)$

29 $2.05\,t + 592\,kg\,(\,kg\,)$

30 $20\,g - 150\,mg\,(\,mg\,)$

31 $36\,kg - 580\,g\,(\,g\,)$

32 $1.5\,t - 590\,kg\,(\,kg\,)$

33 $3.9\,m + 582\,mm\,(\,cm\,)$

34 $0.3\,m - 29.5\,cm\,(\,mm\,)$

Find, in kilograms, the total weight of a bag of flour weighing 1.5 kg, a jar of jam weighing 450 g and a packet of rice weighing 500 g.

The jar of jam weighs $450\,g\ =\ 450 \div 1000\,kg$
$\qquad\qquad\qquad\qquad\quad =\ 0.45\,kg$

The packet of rice weighs $500\,g\ =\ 500 \div 1000\,kg$
$\qquad\qquad\qquad\qquad\qquad =\ 0.5\,kg$

The total weight $=\ (\,1.5 + 0.45 + 0.5\,)\,kg$
$\qquad\qquad\qquad\ =\ 2.45\,kg$

---

35 Subtract 52 kg from 0.8 t, giving your answer in kilograms.

36 Find the difference, in grams, between 5 g and 890 mg.

37 Find the total length, in millimetres, of a piece of wood 82 cm long and another piece of wood 260 mm long.

38 Find the total weight, in kilograms, of 500 g of butter, 2 kg of potatoes and 1.5 kg of flour.

39 One tin of baked beans weighs 220 g.
What is the weight, in kilograms, of ten of these tins?

40 One fence post is 150 cm long.
What length of wood, in metres, is needed to make ten such fence posts?

For question **41** to **50**, choose the most suitable unit for your answer.

41 A wooden vegetable crate and its contents weigh 6.5 kg.
If the crate weighs 1.2 kg what is the weight of its contents?

42 A girl travels to school by walking 450 m to the bus stop and then travelling 1.65 km by bus. The distance she walks after getting off the bus is 130 m.
What distance is her total journey?

43 A man takes three parcels to the Post Office and has them weighed.
One parcel weighs 4 kg 37 g, the weight of the second is 3 kg 982 g and the third one weighs 1 kg 173 g.
What is their total weight?

44 A rectangular field is 947 m long and 581 m wide.
What is the perimeter of the field?
What length of fencing would be needed to go round the field leaving space for two gates, each 3 m wide?

45 Will these three kitchen units fit along a wall that is 300 cm long?

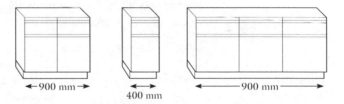

46 Wood is sometimes sold by the 'metric foot'. A metric foot is 30 cm. A man buys a length of wood which is 12 metric feet long. How long is the piece of wood?

47 A freight train has five trucks. Two of them are carrying 15 t 880 kg each. Another has a load of 14 t 700 kg and the last two are each loaded with 24 t 600 kg.
What is the total weight of the contents of the five trucks?
If the weight of each truck is 5 t 260 kg, what is the combined weight of the trucks and their contents?

48 A boy delivers newspapers by bicycle. The weight of the bicycle is 15.8 kg and the boy weighs 51.3 kg. At the beginning of the round the newspapers weigh 9.9 kg.
What is the total weight of the boy and his bicycle loaded with newspapers?
What is the weight when he has delivered half the newspapers?

49 Newton is 5.62 km from Old Town and Old Town is 3.87 km from Castletown.

If a car goes from Newton to Old Town, then to Castletown and finally back to Old Town, how many metres has it travelled?

At the beginning of its journey the car had enough petrol to go 27 km.

At the end of its journey how much further could it go before running out of petrol?

50 A ream of paper (500 sheets) weighs 2.8 kg.

a How much does 1 sheet weigh?

b If a letter weighs more than 60 g, it costs more than one first class stamp to post it. Six of these sheets are put into an envelope. The envelope weighs 8 g.

Will one first class stamp be enough to pay for the postage?

51 A damp patch on a wall rises up the wall by 5 cm every night and recedes by 3 cm during the day. The wall is 3.5 m high and the damp patch starts at the bottom of the wall.

If it carries on like this, how many days will it take the damp patch to reach the top of the wall?

MONEY UNITS

Many countries use units of money that are divided into hundredths. For example

UK	1 pound (£)	= 100 pence (p)
USA	1 dollar ($)	= 100 cents (c)
France	1 franc	= 100 centimes
Germany	1 mark	= 100 pfennigs

EXERCISE 8G

Express each quantity in terms of the unit given in brackets.

1 7 dollars (cents)

2 £6 (pence)

3 8 marks (pfennigs)

4 13 francs (centimes)

5 £2.84 (pence)

6 43 dollars 81 cents (cents)

7 11 marks 3 pfennigs (pfennigs)

8 £6 15 p (pence)

9 $1.30 (cents)

10 7 francs 35 centimes (centimes)

Give 420 p in £.

420 p = £4.20

> We always give pounds to 2 decimal places so we write £4.20 rather than £4.2. Other currencies are written in the same way.

Express each quantity in terms of the unit given in brackets.

11 126 p (£)

12 350 cents (dollars)

13 190 p (£)

14 350 pfennigs (marks)

15 43 dollars 7 cents (dollars)

16 228 p (£)

17 3 marks 47 pfennigs (marks)

18 580 p (£)

19 One tin of baked beans costs 32 p. Find the cost, in pounds, of ten of these tins.

20 Find the total cost, in dollars, of a book costing $4, a pencil costing 30 cents and a magazine costing 75 cents.

21 Find the cost, in pounds, of 20 litres of petrol at 48 p a litre.

22 One can of cola costs 50 pfennigs.
Find the cost, in marks, of twelve such cans.

23 Find the cost of six stamps at 25 p each and five stamps at 19 p each. If you paid for these stamps with a £5 note, how much change would you get?

24 A man takes a parcel to the Post Office. It weighs 3 kg 750 g. The cost of postage is 45 p per 250 g. How much does it cost to post the parcel?

25 A neat pile of six 2 p coins is 10.8 mm high. What is the value of a pile of 2 p coins that is 54 mm high?

MIXED EXERCISE

EXERCISE 8H

Express the given quantity in terms of the units given in brackets.

1 4 km (m)   **4** 250 g (kg)   **7** 1 m 50 cm (m)

2 30 g (kg)   **5** 0.03 km (cm)   **8** 2.8 cm (mm)

3 3.5 m (cm)   **6** 1250 m (km)   **9** 65 g (kg)

10 A tin of meat weighs 429 g. What is the weight, in kilograms, of ten such tins?

11 One pack of fun-size Mars Bars costs £1.20; the pack weighs 90 g and contains 15 bars.

   a How much does each bar cost?

   b How much does each bar weigh?

12 Estimate the lengths of these lines.

   a ————————————

   b ——————————————————

   c ————

   d ————————————————————————

Check your estimates by using your ruler.

PRACTICAL WORK

This drawing shows a woman against a tree.
The woman is 170 cm tall.

a Estimate the height of the tree.

b Use a person or an object (e.g. a door) whose height you know, to estimate the height and width of the main building in your school.

c Explain how you could estimate the length and height of a bridge.

INVESTIGATION

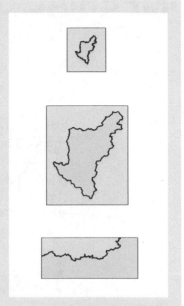

a This is a map of an island. Explain how you could estimate the length of its coastline.

b This is the same island, drawn to a larger scale. Would you get the same answer for the length of its coastline from this drawing?

c This shows the coastline of part of the island drawn to a much larger scale. If you used a map of the whole island drawn with this scale, how would your estimate of the length of the coastline compare with your first estimate?

d Do you think it is possible to measure the length of the coastline exactly? (Think of a bit of coastline you know and imagine measuring a short length of it.)

e Now suppose that you want to measure the length of the table you are sitting at. You could measure it with a ruler. You could measure it with a tape measure marked in centimetres and millimetres. You could measure it with a precision instrument that will read lengths to tenths of a millimetre, or even hundredths of a millimetre. You could measure the length in several different places.
Write down, with reasons, whether it is possible to find the length exactly.
Do you think is possible to give any measurement exactly?

IMPERIAL UNITS

Most measurements in the UK are given in metric units but some Imperial units are still used. For instance, distances on road signs are usually given in miles whereas signposts on foot paths often use kilometres for distances. We need to cope with a mixture of metric and Imperial units. Consider this situation.

David found instructions in a book for making pâpier-maché. Among other things, he needed $1\frac{1}{2}$ lb of flour. He had a 1.5 kg bag of flour and a set of digital scales that weigh in pounds and ounces. David needs to know

- how many ounces there are in one pound so that he can read the scales
- whether he has enough flour, i.e. if 1.5 kg is heavier than $1\frac{1}{2}$ pounds.

To be able to cope with similar situations, we need to be familiar with the Imperial units still in use and to know their approximate values in metric units, and vice-versa.

UNITS OF LENGTH

Miles, yards, feet and inches are the Imperial units of length that are still used. The relationships between them are

> 12 inches (in) = 1 foot (ft)   3 feet = 1 yard (yd)
>
> 1760 yards = 1 mile

EXERCISE 9A

> Express 2 ft 5 in in inches.
>
> $$2\,ft = 2 \times 12\,in$$
> $$= 24\,in$$
> $$\therefore \quad 2\,ft\,5\,in = 24\,in + 5\,in$$
> $$= 29\,in$$
>
> > 2 feet 5 inches is sometimes written as 2′ 5″

Express the given quantity in the unit in brackets.

1 5 ft 8 in (in)   **3** 5 yd 2 ft (ft)

2 4 yd 2 ft (ft)   **4** 10 ft 3 in (in)

5 2 ft 11 in (in)   **7** 9 yd 1 ft (ft)

6 8 ft 4 in (in)   **8** 9 ft 10 in (in)

Express 52″ in feet and inches.

$52'' = 52 \div 12$ ft
$= 4' \; 4''$

There are 4 twelves in 52 with 4 left over.

Express the given quantity in the units in brackets.

9 36 in (ft)   **13** 13 ft (yd and ft)

10 29 in (ft and in)   **14** 75 in (ft and in)

11 86 in (ft and in)   **15** 100 ft (yd and ft)

12 9 ft (yd)   **16** 120 in (ft and in)

17 How many yards are there in

a $1\frac{1}{2}$ miles   **b** $\frac{1}{4}$ mile   **c** $\frac{1}{8}$ mile ?

18 Andy walks to school along a main road. At the start of his journey, a road sign tells him it is $\frac{3}{4}$ mile to the junction where his school is. How many yards does he have to walk?

19 Josh has four cork tiles. Each tile is 14 inches square.
He arranges them in a row to make
a notice board.
How long, in feet and inches, is the board?

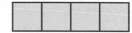

20 Jane has a liquorice lace that is 2 feet long. She cuts it into 6 equal lengths.
How long is each length?

21 Gold coloured chain is sold from a street stall at £1.50 per 4 inches. How much will 2 feet of this chain cost?

22 I passed a sign on a motorway saying 'Road works – 1 mile ahead'. The next sign I passed said 'Road works – 800 yards ahead'. How far apart are the two signs?

23 One lane on a motorway is closed by cones for a distance of 1 mile. The cones are placed 4 feet apart. How many cones are used?

UNITS OF MASS

The Imperial units of weight that are still used are stones, pounds and ounces. Other units of weight that you may still see are hundredweights and tons (not to be confused with tonnes).

The relationships between them are

> 16 ounces (oz) = 1 pound (lb)   14 pounds = 1 stone
>
> 112 pounds = 1 hundredweight (cwt)
>
> 20 hundredweight = 1 ton

EXERCISE 9B

Express the given quantity in terms of the units given in brackets.

1 2 lb 6 oz (oz)   **6** 24 oz (lb and oz)

2 1 lb 12 oz (oz)   **7** 18 oz (lb and oz)

3 4 lb 3 oz (oz)   **8** 36 oz (lb and oz)

4 2 stone 3 lb (lb)   **9** 57 lb (stones and lb)

5 7 stone 6 lb (lb)   **10** 106 lb (stones and lb)

11 3 tons 4 cwt (cwt)   **13** 30 cwt (tons and cwt)

12 1 cwt 50 lb (lb)   **14** 120 lb (cwt and lb)

15 A recipe for Bolognaise sauce needs 12 oz of minced beef for 3 portions of sauce. Amy wants to make 12 portions. What weight of minced beef should she buy?

16 James went on a diet. Before he started, he weighed 10 stone 6 lb. How much did he weigh after losing 10 lb?

17 Joseph bought two bags of bait for fishing. One bag weighed 12 oz and the other bag weighed 9 oz. What is the total weight of this bait in pounds and ounces?

18 The owner of a corner shop buys a 12 lb bag of mint humbugs from which he makes up 32 smaller bags each containing the same weight of humbugs. What is the weight of humbugs in each of these bags?

19 Apples are sold at 64 p per lb. What is the cost of one of these apples weighing 3 ounces?

20 Anthea bought a 56 lb sack of potatoes. She weighed out half of this quantity on her bathroom scales which measure in stones and pounds.
What weight should have shown on the scales?

21 'Pick your own' strawberries are being sold at 64 p per pound. A family picks 8 lb 12 oz of strawberries.
How much will they have to pay for them?

22 A lorry loaded with rubble weighs 12 tons. It delivers 30 cwt of rubble to one site and then 50 cwt of rubble to another site. The tare (empty) weight of the lorry is 5 tons.
What weight of rubble is still in the lorry?

ROUGH EQUIVALENCE BETWEEN METRIC AND IMPERIAL UNITS

When you shop you will find that nearly all prepacked goods (tinned foods, sugar, biscuits, prepacked fruit, etc.) are sold in grams or kilograms and nearly all loose produce (vegetables, fruit, cheese from the delicatessen counter, etc.) is sold in pounds and ounces. It is often useful to be able to convert, roughly, pounds into kilograms and grams into pounds.

At the start of this chapter, David needed to know if 1.5 kg was heavier than $1\frac{1}{2}$ lb. A rough conversion is good enough to answer this question:

> 1 kg is about 2 lb   so that   1 lb is about 500 g

Now we can see that 1.5 kg is more than $1\frac{1}{2}$ lb, so David had enough flour.

> 1 tonne ≈ 1 ton

(1 tonne is very slightly less than 1 ton.)

Rough conversions between metric and Imperial units of length are

> 1 metre is about 1 yd, i.e. 3 feet
>
> 1 mile is about $1\frac{1}{2}$ km   so   1 km is about $\frac{2}{3}$ mile

These rough approximations are good enough for most purposes but there are situations when more accurate conversions are required. For example, if we need to replace a $\frac{1}{2}''$ water pipe, we find that water pipes are now sold with diameters given in millimetres. So that we can decide which is nearest to $\frac{1}{2}''$, it is useful to know that 1 inch ≈ 2.5 cm, i.e 25 mm, so $\frac{1}{2}''$ is about 12.5 mm.

This is easier to remember as

$$\text{4 inches } \approx \text{ 10 cm}$$

Remember that the symbol ≈ means 'is approximately equal to'. Better approximations for other units are

$$\text{1 metre } \approx \text{ 39 inches}$$

(so one metre is slightly longer than one yard)

$$\text{5 miles } \approx \text{ 8 kilometres}$$

(this means that 1 mile is less than 2 km)

$$\text{1 kg } \approx \text{ 2.2 lb}$$

(so 1 kg is a bit more than 2 lb).

EXERCISE 9C   In questions **1** to **10**, write the first unit very roughly in terms of the unit in brackets.

1 3 kg (lb)	**6** 5 m (ft)	**11** 20 ft (m)
2 2 m (ft)	**7** 3.5 kg (lb)	**12** 800 g (lb)
3 4 lb (kg)	**8** 8 ft (m)	**13** 9 lb (kg)
4 9 ft (m)	**9** 250 g (oz)	**14** 50 m (ft)
5 1.5 kg (lb)	**10** 500 g (lb)	**15** 12 kg (lb)

Use the approximation 5 miles $\simeq$ 8 km to convert 60 miles into an approximate number of kilometres.

If 5 miles $\simeq$ 8 km,

then 60 miles $= 5 \times 12$ miles
$$\simeq 8 \times 12 \text{ km} = 96 \text{ km}$$

In questions **16** to **21** use the approximation 5 miles $\simeq$ 8 km to convert the given number of miles into an approximate number of kilometres.

16 10 miles   **18** 15 miles   **20** 75 miles

17 20 miles   **19** 100 miles   **21** 40 miles

22 I buy a 5 lb bag of potatoes and two 1.5 kg bags of flour.
What weight, roughly, in pounds do I have to carry?

23 A window is 6 ft high.
Roughly, what is its height in metres?

24 I have a picture which measures 2 ft by 1 ft. Wood for framing it is sold by the metre.
Roughly, what length of framing, in metres, should I buy?

25 In the supermarket I buy a 4 kg packet of sugar and a 5 lb bag of potatoes.
Which is heavier?

26 In one catalogue a table cloth is described as measuring 4 ft by 8 ft.
In another catalogue a different table cloth is described as measuring 1 m by 2 m.
Which one is bigger?

27 The distance between London and Dover is about 70 miles. The distance between Calais and Paris is about 270 kilometres.
Which is the greater distance?

28 A recipe requires 250 grams of flour.
Roughly, how many ounces is this?

29 An instruction in an old knitting pattern says knit 6 inches. Mary has a tape measure marked only in centimetres.
How many centimetres should she knit?

30 The instructions for repotting a plant say that it should go into a 10 cm pot. The flower pots that Tom has in his shed are marked 3 in, 4 in and 5 in.
Which one should he use?

31 Peter Stuart wishes to extend his central heating which was installed several years ago using 1 in and $\frac{1}{2}$ in diameter copper tubing. The only new piping he can buy has diameters of 10 mm, 15 mm, 20 mm or 25 mm.

Use the approximation 1 in $\approx$ 2.54 cm to determine which piping he should buy that would be nearest to

a the 1 in pipes   **b** the $\frac{1}{2}$ in pipes.

32 A carpenter wishes to replace a 6 in floorboard. The only sizes available are metric and have widths of 12 cm, 15 cm, 18 cm and 20 cm.

Which one should he buy?

33 Eddy knows his height is 4 ft 5 in. He needs to fill in a passport application form and has to give his height in metres.

What should he enter for his height?

34 The doctor tells Mr Brown that he needs to lose 10 kilograms in weight. Mr Brown's scales at home show his weight now as 15 stone 6 lb.

What will his scales show when he has lost the required weight?

35 A shop sells material at £10.50 per metre while the same material is sold in the local market at £9 per yard.

Which is cheaper?

36 Arrange these weights in order of size with the lightest first.

$$2\,\text{oz}, \quad 50\,\text{g}, \quad 0.04\,\text{kg}, \quad \tfrac{1}{5}\,\text{lb}$$

37 Arrange these lengths in order of size with the longest first.

$$25\,\text{cm}, \quad 8\,\text{inches}, \quad 25\,\text{mm}, \quad 1\tfrac{1}{8}\,\text{inches}$$

INVESTIGATION

There are other Imperial units that have specialised uses, for example, furlongs are used to measure distances in horse racing, fathoms are used to measure the depth of water.

a Use reference materials to find out the relationships between these units and the more common Imperial units of length.

b Find out as much as you can about other Imperial units of distance and weight.

c Nautical miles are used to measure distances at sea. Find out what you can about nautical miles, including the rough equivalence of 1 nautical mile in miles and in kilometres.

**PRACTICAL
WORK**

This is a group exercise.
A group of year 7 pupils were asked to write down their heights and weights on sheets of paper which were then gathered in.
This is a list of *exactly* what was written down.

Height	Weight
141 cm	35 kg
138 cm	4 stone
1.8 m	6.26 stone
4 feet 5 inches	4 kg
52 feet	6 stone
5 foot 4	8 stone
1 metre 53	$7\frac{1}{2}$ stone
1 metre 41 cm	28.0 kg
141 cm	5 stone 4 pounds
4 feet 7 inches	32 kg

a This group of children used a mixture of units. Some of the entries are unbelievable.
Which are they?
Discuss what the reasons might be for these unbelievable entries.

b Find out how your group know their heights and weights; each of you write down your own height and weight on a piece of paper.
Use whatever unit you know them in, and do not write your name on it.
Collect in the pieces of paper and write out a list like the one above.

c What official forms do you know about that ask for height?
What unit is required?

d Write down your own height and weight in both metric and Imperial units.

SUMMARY 2

FRACTIONS

The bottom number (denominator) of a fraction tells you how many equal-sized parts the whole quantity is divided into. The top number (numerator) tells you how many of these parts are being considered,

e.g. $\frac{2}{5}$ of a bar of chocolate means 2 out of the 5 equal-sized parts of the bar of chocolate.

Equivalent fractions are formed by multiplying or by dividing the top and the bottom of a fraction by the same number,

e.g. $\frac{1}{2} = \frac{3}{6}$ and $\frac{5}{10} = \frac{1}{2}$

PERCENTAGES

A percentage of a quantity describes how many of the 100 equal-sized parts of the quantity are being considered,

e.g. 20% of an apple means 20 out of the 100 equal-sized parts of the apple,

i.e. $20\% = \frac{20}{100}$

DECIMALS

In decimal notation, numbers to the right of the decimal point represent tenths, hundredths, thousandths, ...,

e.g. $0.534 = \frac{5}{10} + \frac{3}{100} + \frac{4}{1000}$

CHANGING DECIMALS TO FRACTIONS

A decimal can be changed to a fraction by using the number after the decimal point as the numerator and 10, 100, 1000, ... as the denominator,

e.g. $0.56 = \frac{56}{100} = \frac{14}{25}$

$0.025 = \frac{25}{1000} = \frac{1}{40}$

ADDITION AND SUBTRACTION OF FRACTIONS

Fractions can be added or subtracted when they have the same denominator,

e.g. to add $\frac{1}{2}$ to $\frac{1}{3}$ we must first change them into equivalent fractions with the same denominators,

i.e. $\frac{1}{2} + \frac{1}{3} = \frac{1}{2} \times \frac{3}{3} + \frac{1}{3} \times \frac{2}{2} = \frac{3}{6} + \frac{2}{6} = \frac{5}{6}$

ADDITION AND SUBTRACTION OF DECIMALS	Decimals can be added or subtracted using the same methods as for whole numbers, provided the decimal points are placed in line,

e.g.   to add 12.56 and 7.9, we can write them as

$$\begin{array}{r} 12.56 \\ +\ \ 7.9 \\ \hline 20.46 \end{array}$$

MULTIPLYING AND DIVIDING DECIMALS BY 10, 100, 1000, ...	To multiply a decimal by 10, 100, 1000, ... we move the point 1, 2, 3, ... places to the right,

e.g. $2.56 \times 10 = 25.6,$ and $2.56 \times 1000 = 2560(.0\,)$

To divide a decimal by 10, 100, 1000, ... we move the point 1, 2, 3, ... places to the left,

e.g. $2.56 \div 10 = 0.256,$ and $2.56 \div 1000 = 0.002\,56$

MULTIPLYING DECIMALS	To multiply decimals without using a calculator, first ignore the decimal point and multiply the numbers. Then add together the number of decimal places in each of the decimals being multiplied; this gives the number of decimal places in the answer,

e.g. $2.5 \times 0.4 = 1.00 = 1$ $(25 \times 4 = 100)$
$$[(1) + (1) = (2)]$$

DIVIDING BY A DECIMAL	To divide by a decimal, move the point in *both* numbers to the right until the number we are dividing by is a whole number,

e.g. $2.56 \div 0.4 = 25.6 \div 4$

Now we can use ordinary division, keeping the decimal point in the same place,

e.g. $25.6 \div 4 = 6.4$ $4\overline{)25.6}$ with 6.4

CORRECTING DECIMALS	To round (i.e. to correct) a number to a specified number of decimal places, look at the figure in the next decimal place: if it is 5 or more, add 1 to the previous figure, otherwise leave the previous figure as it is,

e.g.   to write 2.564 correct to 2 decimal places, we have

    $2.56\,|\,4 = 2.56$ correct to 2 decimal places,

and to write 2.564 correct to 1 decimal place, we have

    $2.5\,|\,64 = 2.6$ correct to 1 decimal place.

CHANGING A FRACTION TO A DECIMAL

Fractions can be changed to decimal notation by dividing the bottom number into the top number,

e.g. $\frac{3}{8} = 3 \div 8 = 0.375$

METRIC UNITS

The metric units of length in common use are the kilometre, the metre, the centimetre and the millimetre, where

$$1\,\text{km} = 1000\,\text{m}, \qquad 1\,\text{m} = 100\,\text{cm}, \qquad 1\,\text{cm} = 10\,\text{mm}$$

The metric units of weight are the tonne, the kilogram, the gram and the milligram, where

$$1\,\text{tonne} = 1000\,\text{kg}, \qquad 1\,\text{kg} = 1000\,\text{g}, \qquad 1\,\text{g} = 1000\,\text{mg}$$

IMPERIAL UNITS

The Imperial units of length in common use are the mile, the yard, the foot and the inch, where

$$1\,\text{mile} = 1760\,\text{yards}, \qquad 1\,\text{yard} = 3\,\text{feet}, \qquad 1\,\text{foot} = 12\,\text{inches}$$

The Imperial units of weight still in common use are the ton, the stone, the pound and ounce, where

$$1\,\text{ton} = 2240\,\text{lb}, \qquad 1\,\text{stone} = 14\,\text{lb}, \qquad 1\,\text{lb} = 16\,\text{ounces}$$

EQUIVALENCE BETWEEN METRIC AND IMPERIAL UNITS

For a rough conversion between the two sets of units, use

$$1\,\text{mile} \simeq \tfrac{1}{2}\,\text{km}, \qquad 1\,\text{yard} \simeq 1\,\text{m},$$
$$1\,\text{kg} \simeq 2\,\text{lb}, \qquad 1\,\text{tonne} \simeq 1\,\text{ton}$$

For a better approximation use

$$5\,\text{miles} \simeq 8\,\text{km}, \qquad 1\,\text{inch} \simeq 2.5\,\text{cm}, \qquad 1\,\text{kg} \simeq 2.2\,\text{lb}$$

REVISION EXERCISE 2.1 (Chapters 5 and 6)

1 Change each fraction into an equivalent fraction with a denominator of 24

  a $\frac{1}{3}$ **b** $\frac{3}{4}$ **c** $\frac{1}{6}$ **d** $\frac{7}{12}$ **e** $\frac{7}{8}$

2 Express as a decimal   **a** $\frac{7}{10}$ **b** $\frac{7}{100}$ **c** $\frac{90}{100}$ **d** $\frac{31}{100}$

3 In a golf club $\frac{4}{5}$ of the members are men.

  a What percentage is this?

  b If there are 450 members, how many of them are men?

4 Express as a fraction in its lowest terms

    a 0.75   **b** 0.2   **c** 0.05   **d** 0.15

5 Due to a reduction in orders the workforce at a factory was reduced by 12%. What fraction is this?

6 Find   **a** $\frac{1}{5} + \frac{1}{10}$ **b** $\frac{2}{3} + \frac{1}{7}$ **c** $1\frac{1}{3} + \frac{3}{4}$ **d** $4\frac{7}{12} - 2\frac{1}{6}$

7 Evaluate   **a** $0.05 + 2.17$ **b** $31.7 + 5.34$ **c** $3.02 + 0.9$

8 Find   **a** $\frac{2}{3} - \frac{1}{5}$ **b** $1\frac{1}{4} - \frac{1}{2}$ **c** $9\frac{1}{3} + 4\frac{5}{6}$

9 Find   **a** $7.8 - 0.36$ **b** $3.59 - 1.2$

10 In 1970 about 3 households in every 10 did not have a car. What percentage of households

    a did not have a car   **b** had a car?

REVISION EXERCISE 2.2 (Chapters 7, 8 and 9)

1 Find, without using a calculator,

    a 0.25×1.2 **c** $8.4 \div 0.02$

    b 12.03×0.5 **d** $45.6 \div 20$

2 a Estimate the value of $13.23 \div 21$, and then use a calculator to find its exact value.

    b Use a calculator to find $53.4 \div 9$ correct to 2 decimal places.

3 One box of biscuits costs £1.39. Find the cost of 10 of these boxes.

4 Express the given quantity in terms of the unit in brackets.

    a 480 cm (m)   **b** 2.5 kg (g)   **c** 0.05 g (mg)

5 Express the given quantity in terms of the unit in brackets.

    a 2 feet (inches)   **b** 2 stone (lb)   **c** $1\frac{1}{2}$ lb (oz)

6 How many inches, roughly, are equivalent to 60 cm?

7 Place these lengths in order of size with the smallest first.

    45 inches,   573 mm,   0.78 km,   500 yd

8 Measure this line in millimetres.

_____

What is the length of one quarter of this line?

9 Use this drawing to estimate the height, in millimetres, of the back of the lectern. The man is about 190 cm tall.

10 How far, roughly, is

a 48 miles in kilometres

b 100 kilometres in miles.

REVISION EXERCISE 2.3 (Chapters 5 to 9)

1 a Simplify the fractions **i** $\frac{36}{48}$ **ii** $\frac{15}{25}$ **iii** $\frac{54}{90}$

b Find **i** $\frac{4}{5} + \frac{3}{4}$ **ii** $1\frac{1}{2} - \frac{2}{3}$ **iii** $1\frac{1}{2} + \frac{2}{3}$

2 Express **a** 41 centimetres in millimetres

b 3516 millimetres in metres

c 7526 kg in tonnes

3 An alloy is 65% copper and the rest is tin.

a What fraction of the alloy is copper?

b Express the part that is tin as
i a percentage **ii** a decimal **iii** a fraction in its lowest terms.

4 A pack of eight cans of cola costs £1.76.
Find the cost of one of these cans.

5 The diagram shows part of the display on a voltmeter.

What is the reading?

6 Place these weights in order of size, with the smallest first.

2 lb, 3 kg, $\frac{1}{2}$ stone, 500 g, 40 oz

7 a Use your calculator to find **i** 8.76×5.2 **ii** 9.37×8.04

b Use your calculator to find, correct to 2 decimal places

i $4.37 \div 7.7$ **ii** $2193 \div 52$ **iii** $4.56 \div 6$

8 A piece of material is $2\frac{1}{8}$ yards long. Carl cuts off a piece $1\frac{1}{3}$ yards long.
What length is left?

9 One paving stone is 382 mm long. Estimate, in metres, the length of 37 of these paving stones placed end to end.

10 This piece of wood is 3.5 m long.

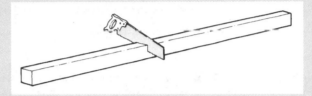

Liam cuts off a piece 1465 mm long.
How much is left?

**REVISION
EXERCISE 2.4
(Chapters 1 to 9)**

1 What is the difference between 173 and 294?

2 Find, without using a calculator, the value of

a 145×20 **b** 803×40 **c** $210 \div 30$

3 Which is the heavier, a 2 lb bag of apples or a 1 kg bag of oranges?

4 Pete writes down his height as 131 cm. Joanne writes down her height as 4 ft 5 in. Who is the taller and by how much?

5

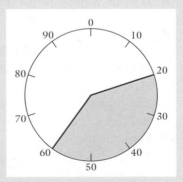

Express the part of the circle that is shaded as

a a percentage **c** a fraction in its lowest terms.

b a decimal

6 To 0.74 add the sum of 4.089 and 0.77.

7 The height of a pile of twenty 20 p coins is 34 mm.
How thick is one coin?

8 Find, without using a calculator,   **a** $\frac{3}{4} + \frac{5}{12}$ **b** $\frac{9}{10} - \frac{2}{15}$

9 Express the given quantity in the unit in brackets.

   a 4 ft 7 in (in)   **b** 7 yd 1 ft (ft)

10 Eve wants to buy some 6 inch plant pots. At the garden centre she sees pots marked 10 cm, 15 cm, 20 cm and 25 cm.
Which size should she buy?

**REVISION
EXERCISE 2.5
(Chapters 1 to 9)**

1 **a** Sally uses 3 kg of potatoes a day.
How many days will a 50 kg bag last?

   b Jean loads a ream of paper (500 sheets) into the tray of her photocopier. She has 23 sheets to copy and needs 20 copies of each sheet.
How many sheets of paper remain unused?

2 **a** Find, giving the remainder if there is one,
   i $429 \div 37$ **ii** $2193 \div 53$

   b Find
   i $3\frac{7}{12} + 1\frac{1}{4}$ **ii** $5\frac{1}{4} - 2\frac{5}{8}$ **iii** $3 \times 5 - 4 \times 2 - 16 \div 4$

3 Find the sum of 3.07 and 0.76 and subtract the result from 7.56.

4 A rectangular table top, measuring 96 cm by 66 cm, is to be tiled with square ceramic tiles.
Find the size of the largest tile that can be used if there is to be no cutting.

5 Jim cleared 2.50 metres in a high-jump competition.
Find the height of the bar in feet and inches given that
1 inch $\simeq$ 2.54 cm.

6 Amy measured the length of a screw with a ruler. The reading was $1\frac{1}{4}$ inches. Write this measurement in decimal form.

7 Which is the larger   **a** 10 inches or 20 cm   **b** 3 tons or 3 tonnes?

8 Find the difference between **a** 8 g and 650 mg **b** 0.5 t and 64 kg.
(Give each answer in the smaller unit.)

9 A piece of wood is 9 ft 4 in long. It is cut into 8 equal lengths.
How long is each piece ?

10 The bar chart shows the number of £5, £10, £20 and £50 notes
paid into a bank account by a businessman on a visit to the bank.

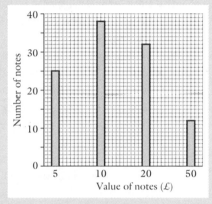

a How many £5 notes were paid in ?

b What was the value of all the £10 notes paid in ?

c How many notes did the bank teller have to count ?

d What was the total value of the notes that were paid in ?

INTRODUCING GEOMETRY

10

You have arrived in a strange town and wish to visit the castle. There are two things you need to know. In which direction is the castle? How far is it?

- The directions you are given could well be: Go to the end of the street, take the first right and then the second left.
- For an aeroplane flying from London to Los Angeles the directions would not be given in this crude way. Very sensitive instruments give accurate directions which are checked and rechecked at regular intervals.
- When the Channel Tunnel was built it was necessary to measure direction very accurately. Discuss the consequences of poor measurements.

Can you think of other situations where we have to measure directions accurately?

The examples above show that we need to give and measure directions accurately.

To describe a direction we measure the amount of turning from some fixed direction. This fixed direction could be the main street in a town, the direction of true north or the position of the minute hand of a clock at a particular time. Having decided on the fixed direction we measure the amount of turn needed to give the direction we seek.

FRACTIONS OF A REVOLUTION

When the seconds hand of a clock starts at 12 and moves round until it stops at 12 again it has gone through one complete turn.

One complete turn is called a revolution.

When the seconds hand starts at 12 and stops at 3 it has turned through $\frac{1}{4}$ of a revolution.

EXERCISE 10A

What fraction of a revolution does the seconds hand of a clock turn through when

a it starts at 3 and stops at 12

b it starts at 4 and stops at 8?

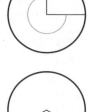

a If the seconds hand starts at 3 and stops at 12 it turns through $\frac{3}{4}$ of a revolution.

b If the seconds hand starts at 4 and stops at 8 it turns through $\frac{1}{3}$ of a revolution.

What fraction of a revolution does the seconds hand of a clock turn through when it

1 starts at 12 and stops at 9

2 starts at 12 and stops at 6

3 starts at 6 and stops at 9

4 starts at 3 and stops at 9

5 starts at 9 and stops at 12

6 starts at 1 and stops at 7

7 starts at 5 and stops at 11

8 starts at 10 and stops at 4

9 starts at 8 and stops at 8

10 starts at 8 and stops at 11

11 starts at 10 and stops at 2

12 starts at 12 and stops at 4

13 starts at 8 and stops at 5

14 starts at 5 and stops at 2?

Where does the hand stop if it starts at 12 and turns through $\frac{1}{4}$ of a revolution?

It stops at 3.

Where does the hand stop if

15 it starts at 12 and turns through $\frac{1}{2}$ a turn

16 it starts at 12 and turns through $\frac{3}{4}$ of a turn

17 it starts at 6 and turns through $\frac{1}{4}$ of a turn

18 it starts at 9 and turns through $\frac{1}{2}$ a turn

19 it starts at 6 and turns through a complete revolution

20 it starts at 9 and turns through $\frac{3}{4}$ of a revolution

21 it starts at 12 and turns through $\frac{1}{3}$ of a revolution

22 it starts at 12 and turns through $\frac{2}{3}$ of a revolution

23 it starts at 9 and turns through a complete revolution?

COMPASS DIRECTIONS

The four main compass directions are north, south, east and west.

If you stand facing north and turn clockwise through $\frac{1}{2}$ a revolution you are then facing south.

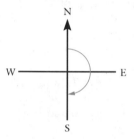

EXERCISE 10B

1 If you stand facing west and turn anticlockwise through $\frac{3}{4}$ of a revolution, in which direction are you facing?

2 If you stand facing south and turn clockwise through $\frac{1}{4}$ of a revolution, in which direction are you facing?

3 If you stand facing north and turn, in either direction, through a complete revolution, in which direction are you facing?

4 If you stand facing west and turn through $\frac{1}{2}$ a revolution, in which direction are you facing? Does it matter if you turn clockwise or anticlockwise?

5 If you stand facing south and turn through $1\frac{1}{2}$ revolutions, in which direction are you facing?

6 If you stand facing west and turn clockwise to face south what part of a revolution have you turned through?

7 If you stand facing east and turn to face west what part of a revolution have you turned through?

ANGLES

When the hand of a clock moves from one position to another it has turned through an angle.

RIGHT ANGLES

A quarter of a revolution is called a *right angle*.

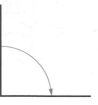

Half a revolution is two right angles.

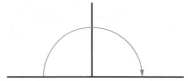

EXERCISE 10C

> How many right angles does the seconds hand of a clock turn through when it starts at 3 and stops at 12 ?
>
> It turns through three right angles.

How many right angles does the seconds hand of a clock turn through when it

1   starts at 6 and stops at 9

2   starts at 3 and stops at 9

3   starts at 12 and stops at 9

4   starts at 3 and stops at 6

5   starts at 12 and stops at 12

6   starts at 8 and stops at 2

7   starts at 9 and stops at 6

8   starts at 7 and stops at 7 ?

How many right angles do you turn through if you

9   face north and turn clockwise to face south

10   face west and turn clockwise to face north

11   face south and turn clockwise to face west

12   face north and turn anticlockwise to face east

13   face north and turn to face north again ?

ACUTE, OBTUSE AND REFLEX ANGLES

Any angle that is smaller than a right angle is called an *acute angle*.

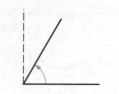

Any angle that is greater than one right angle and less than two right angles is called an *obtuse angle*.

Any angle that is greater than two right angles is called a *reflex angle*.

EXERCISE 10D   What type is each of the following angles?

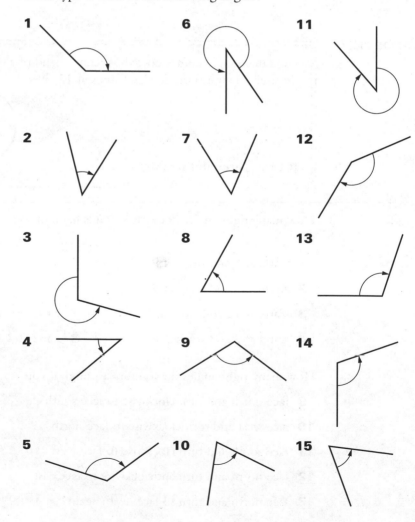

DEGREES

One complete revolution is divided into 360 parts. Each part is called a *degree*. 360 degrees is written 360°.

360 seems a strange number of parts to have in a revolution but it is a good number because so many whole numbers divide into it exactly. This means that there are many fractions of a revolution that can be expressed as an exact number of degrees.

It is possible that the number 360 was chosen because thousands of years ago the Babylonians believed that the sun took 360 days to complete one circuit of the earth i.e. to make a complete turn.

EXERCISE 10E

1 How many degrees are there in half a revolution?

2 How many degrees are there in one right angle?

3 How many degrees are there in three right angles?

How many degrees has the seconds hand of a clock turned through when it moves from 6 to 9?

It has turned through 90°.

How many degrees has the seconds hand of a clock turned through when it moves from

4 12 to 6	**8** 9 to 6	**12** 8 to 5
5 3 to 6	**9** 2 to 5	**13** 4 to 10
6 6 to 3	**10** 7 to 10	**14** 5 to 8
7 9 to 3	**11** 1 to 10	**15** 6 to 12?

How many degrees has the seconds hand of a clock turned through when it moves from 6 to 8?

> When the hand moves from 6 to 8 it turns through $\frac{2}{12}$ of a revolution.

The hand turns through $\frac{2}{12}$ of 360°, i.e. 60°

How many degrees has the seconds hand of a clock turned through when it moves from

16 8 to 9

17 10 to halfway between 11 and 12

18 6 to 10

19 1 to 3

20 3 to halfway between 4 and 5

21 4 to 5

22 7 to 11

23 5 to 6

24 7 to 9

25 11 to 3

26 3 to 10

27 2 to 8

28 10 to 8

29 12 to 11

30 9 to 2

31 8 to 3

32 7 to 5

33 10 to 5

34 11 to 4

35 2 to 9?

36 You wish to divide a complete turn into an exact number of parts.
Is 360 a good number?
Try dividing 360 by each of the whole numbers from 1 to 12.
Now do the same for 100.
Which is the better number of degrees to have in one complete turn, 100 or 360?
Give reasons for your choice.

USING A PROTRACTOR TO MEASURE ANGLES

A protractor looks like this:

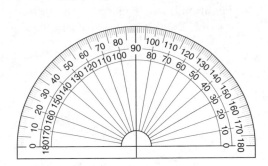

It has a straight line at or near the straight edge. This line is called the *base line*. The *centre* of the base line is marked.

The protractor has two scales, an inside one and an outside one.

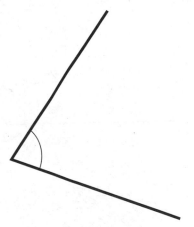

To measure the size of this angle, first decide whether it is acute or obtuse.

This is an acute angle because it is *less* than 90°.

Next place the protractor on the angle as shown.

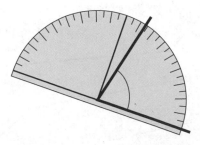

One arm of the angle is on the base line.

The vertex (point) of the angle is at the centre of the base line.

Choose the scale that starts at 0° on the arm on the base line. Read off the number where the other arm cuts this scale.

Check with the earlier decsion about the size of the angle to make sure that you have chosen the right scale.

EXERCISE 10F Measure the following angles (if necessary, turn the page to a convenient position).

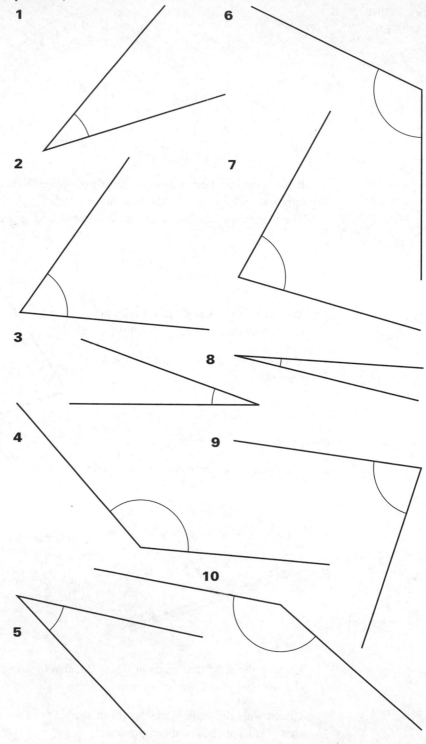

Find the size of the angle marked *p*.

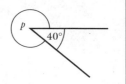

Angle *p* and 40° together make 360°
So angle *p* is 360° − 40° = 320°.

In questions **11** to **15** write down the size of the angle marked with a letter.

11

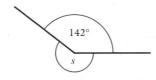

12

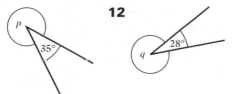

13

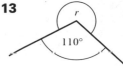

14

15

Measure the following angle.

This is a reflex angle and it is bigger than 3 right angles, i.e. it is greater than 270°.

To find this angle, we need to measure the smaller angle, marked *p*.

Angle *p* is 68° so the reflex angle is 360° − 68° = 292°.

Measure the following angles.

16

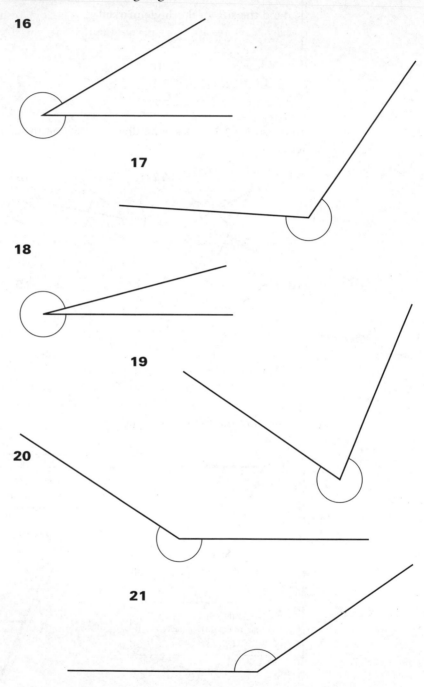

17

18

19

20

21

22 Draw a reflex angle. Now find its size.
Change books with your neighbour and each check the other's measurement.

EXERCISE 10G

This exercise contains mixed questions on angles. Use a clock diagram to draw the angle that the *minute* hand of a clock turns through in the following times. In each question write down the size of the angle in degrees.

1 5 minutes **3** 15 minutes **5** 25 minutes

2 10 minutes **4** 20 minutes **6** 30 minutes

The seconds hand of a clock starts at 12. Which number is it pointing to when it has turned through an angle of

7 90° **11** 150° **15** 420° **19** 540°

8 60° **12** 270° **16** 180° **20** 240°

9 120° **13** 30° **17** 450° **21** 390°

10 360° **14** 300° **18** 210° **22** 720° ?

If you stand facing north and turn clockwise, draw a sketch to show roughly the direction in which you are facing if you turn through 60°.

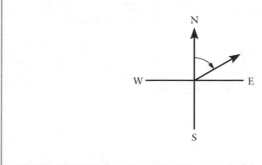

If you start by facing north and turn clockwise, draw a sketch to show roughly the direction in which you are facing if you turn through the following angles.

23 45° **26** 50° **29** 20° **32** 10°

24 70° **27** 200° **30** 100° **33** 80°

25 120° **28** 300° **31** 270° **34** 250°

Estimate the size, in degrees, of each of the following angles.

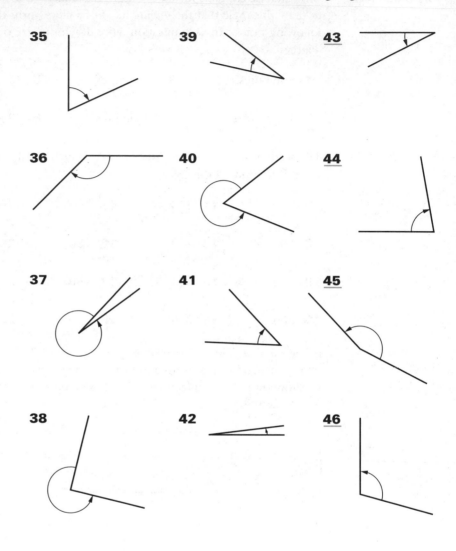

35

36

37

38

39

40

41

42

43

44

45

46

Draw the following angles as well as you can by estimating, i.e. without using a protractor. Use a clockface if it helps. Then measure your angles with a protractor.

47 45°	**50** 30°	**53** 150°	**56** 20°	**59** 330°
48 90°	**51** 60°	**54** 200°	**57** 5°	**60** 95°
49 120°	**52** 10°	**55** 290°	**58** 170°	**61** 250°

DRAWING ANGLES USING A PROTRACTOR

To draw an angle of 120° start by drawing one arm and mark the corner.

The corner, or point of an angle, is called the *vertex*.

Place your protractor as shown in the diagram. Make sure that the vertex is at the centre of the base line.

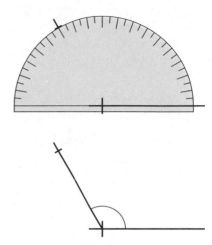

Choose the scale that starts at 0° on your drawn line and mark the paper next to the 120° mark on the scale.

Remove the protractor and join your mark to the vertex.

Now look at your angle: does it look the right size?

EXERCISE 10H

Use your protractor to draw the following angles accurately.

1 25°	**4** 160°	**7** 110°	**10** 125°	**13** 105°
2 37°	**5** 83°	**8** 49°	**11** 175°	**14** 136°
3 55°	**6** 15°	**9** 65°	**12** 72°	**15** 85°

Change books with your neighbour and measure each other's angles as a check on accuracy.

EXERCISE 10I

In questions **1** and **2** first measure the angle marked *r*. Then estimate the size of the angle marked *s*. Check your estimate by measuring angle *s*.

1

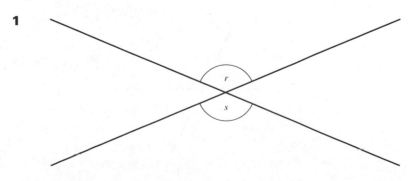

2

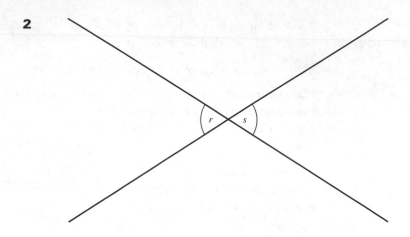

3 Draw some more similar diagrams and repeat questions **1** and **2**.

In each of the following questions, write down the size of the angle marked *t*, without measuring it.

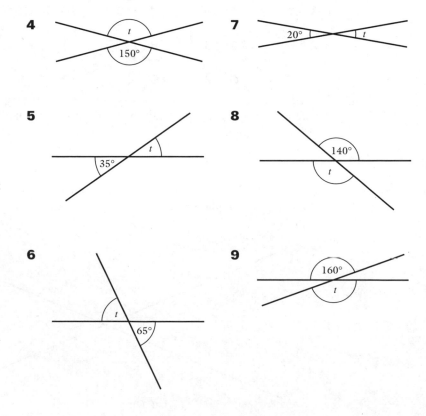

4

150°

7

20°

5

35°

8

140°

6

65°

9

160°

VERTICALLY OPPOSITE ANGLES

When two straight lines cross, four angles are formed.

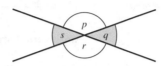

The two angles that are opposite each other are called *vertically opposite angles*. After working through the last exercise you should now be convinced that

> vertically opposite angles are equal

i.e. $p = r$ and $s = q$

ANGLES ON A STRAIGHT LINE

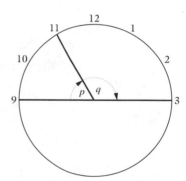

The seconds hand of a clock starts at 9 and stops at 11 and then starts again and finally stops at 3.

Altogether the seconds hand has turned through half a revolution, so $p + q = 180°$.

EXERCISE 10J

1 Draw a diagram showing the two angles that you turn through if you start by facing north and then turn clockwise through $60°$, stop for a moment and then continue turning until you are facing south. What is the sum of these two angles?

2 Draw a clock diagram to show the two angles turned through by the seconds hand if it is started at 2, stopped at 6, started again and finally stopped at 8. What is the sum of these two angles?

3 Draw an angle of $180°$, without using your protractor.

4 Darren asks you what an angle of $180°$ is.
Write down how you would answer Darren.

**SUPPLEMENTARY
ANGLES**

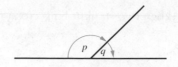

> Angles on a straight line add up to 180°.

> Two angles that add up to 180 ° are called supplementary angles.

EXERCISE 10K

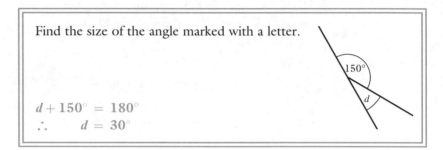

Find the size of the angle marked with a letter.

$d + 150° = 180°$
$\therefore \qquad d = 30°$

In questions **1** to **12** calculate the size of the angle marked with a letter.

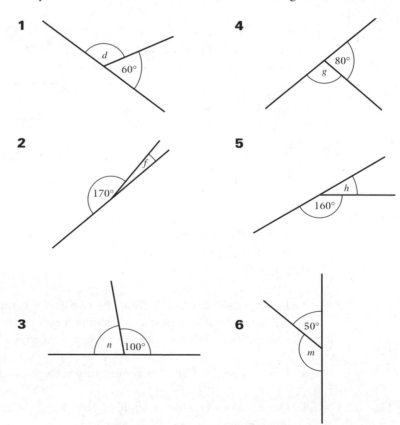

1

d

$60°$

2

$170°$

f

3

n $\quad 100°$

4

$80°$

g

5

h

$160°$

6

$50°$

m

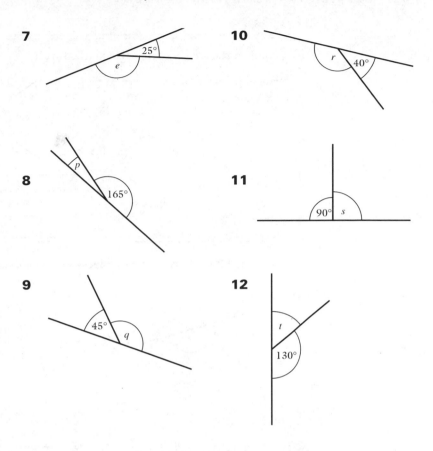

7

25°

e

10

r 40°

8

p

165°

11

90° s

9

45°

q

12

t

130°

In questions **13** to **18** write down the pairs of angles that are
supplementary.

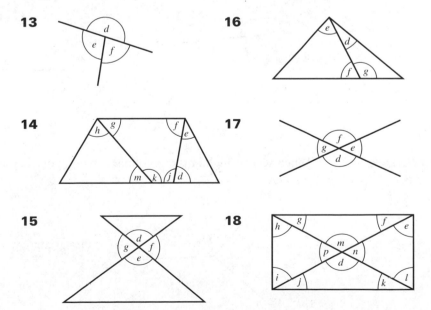

13

d
e f

16

e
d
f g

14

h g f e
m k j d

17

f
g e
d

15

d
g f
e

18

h g f e
m
p n
d
i j k l

Calculate the size of each angle marked with a letter.

d and 70° are equal
(they are vertically opposite)
∴ *d* = 70°
e and 70° add up 180°
(they are angles on a straight line)
∴ *e* = 110°
f and *e* are equal
(they are vertically opposite)
∴ *f* = 110°

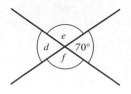

Notice that reasons are given for statements made about angles.

In questions **19** to **24** calculate the sizes of the angles marked with a letter.

19

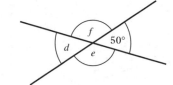

22

20

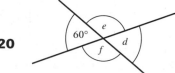

23

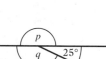

21

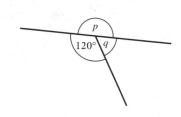

24

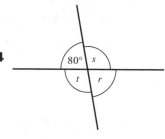

ANGLES AT A POINT

When several angles make a complete revolution they are called *angles at a point*.

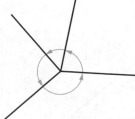

Angles at a point add up to 360°.

EXERCISE 10L

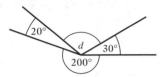

Calculate the size of each angle marked with a letter.

The three given angles add up to 250°.

$$\therefore \quad d = 360° - 250°$$
$$d = 110°$$

$$\begin{array}{r} 30 \\ 200 \\ + \ 20 \\ \hline 250 \end{array}$$

Find the size of each angle marked with a letter. Show your working and give reasons.

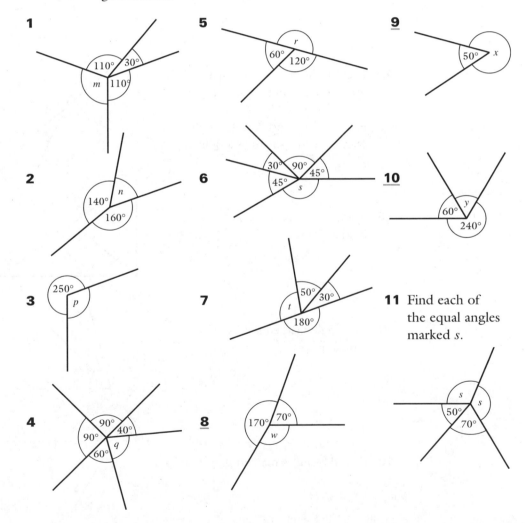

1

5

9

2

6

10

3

7

11 Find each of the equal angles marked *s*.

4

8

12 Find each of the equal angles marked *d*.

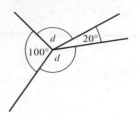

13 Each of the equal angles marked *p* is 25°.
Find the reflex angle *q*.

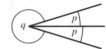

14 The angle marked *f* is twice the angle
marked *g*.
Find angles *f* and *g*.

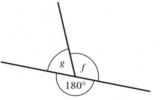

15 Each of the equal angles marked *d* is
30°. Angle *d* and angle *e* are
supplementary.
Find angles *e* and *f*. (An angle
marked with a square is a right angle.)

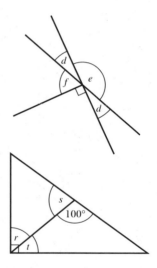

16 Angle *s* is twice angle *t*.
Find angle *r*.

17 The angle marked *d* is 70°.
Find angle *e*.

18 Find the angles marked *p*, *q*, *r* and *s*.

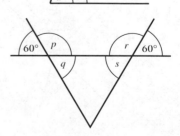

1 What angle does the minute hand of a clock turn through when it moves from 1 to 9?

2 Draw an angle of 50°.

3 Estimate the size of this angle.

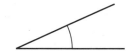

4 Write down the size of the angle marked *p*.

5 Write down the size of the angle marked *s*.

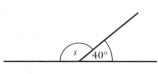

6 Find each of the equal angles marked *e*.

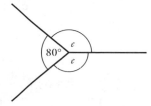

PRACTICAL
WORK

This is an exercise for two people. You need a good map and a protractor.

Toss a coin to see who takes the first turn.

One player finds two places on the map. This player shows the other player the position of one place on the map and gives the direction of the second place from the first by estimating the angle that must be turned through clockwise from north. If the second player finds this place within 10 seconds, he has won and it is his turn. Otherwise the first player has won and gets another turn. Play as many times as you wish.

The winner is the player with the most successes.

Any disputes about the direction given are solved by measuring the actual angle with a protractor. Directions within 10° are acceptable.

If 10 seconds is not long enough, increase the time allowed to 15 seconds.

INVESTIGATION

a How do you find the angle turned through by the hour hand of a clock in a given time?

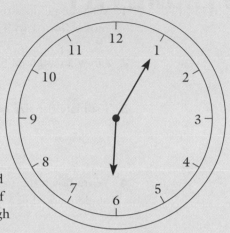

Start by finding the angle turned through in 12 hours, then the angle for any other complete number of hours.
Next find the angle turned through for any fraction of an hour and, lastly, through a number of minutes.

b Extend your investigation to the minute hand and the seconds hand.

c How do you find the angle between the hands of a clock at any time?
Start with times that give you the angles that are easiest to find. Remember that at 4.30 the minute hand will point to the 6 and the hour hand will be exactly half way between 4 and 5.

d Find out how many times there are in a day when the angle between the hour hand and the minute hand has a particular value – say 90°, 180° or 120°.

e What happens if the clock loses 10 minutes each hour?
How many degrees would the minute hand turn through in 1 hour, or 15 minutes or any other time?

f What happens if the clock gains 5 minutes every hour?

SYMMETRY

11

To most people symmetrical shapes are pleasing to the eye.
These shapes are all around us, whether they are man-made things like
Concorde or a suspension bridge, or shapes that occur in nature such as a
snowflake or an open flower.

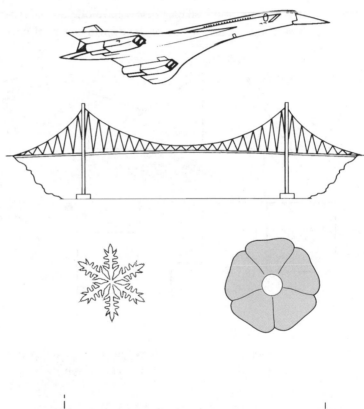

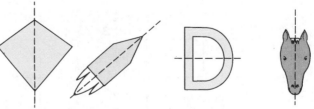

The four shapes above are *symmetrical*. If they were folded along the
broken line, one half of the drawing would fit exactly over the other half.

Fold a piece of paper in half and cut a shape from the folded edge. When unfolded, the resulting shape is symmetrical. The fold line is the *axis of symmetry*.

EXERCISE 11A

Some of the shapes below have one axis of symmetry and some have none. State which of the drawings **1** to **6** have an axis of symmetry.

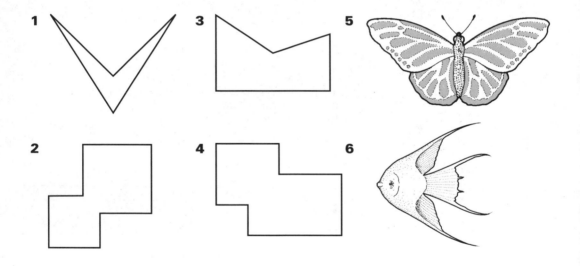

Copy the following drawings on squared paper and complete them so that the broken line is the axis of symmetry.

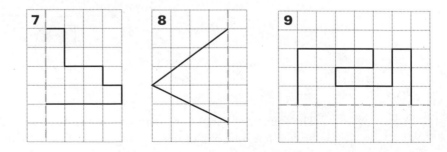

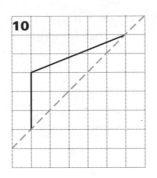

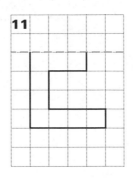

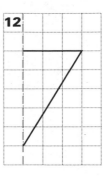

13 Copy and complete each letter so that the dashed lines are lines of symmetry.

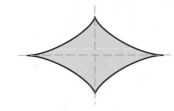

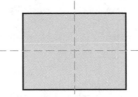

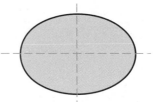

**TWO AXES OF
SYMMETRY**

In these shapes there are two lines along which it is possible to fold the paper so that one half fits exactly over the other half.

Fold a piece of paper twice, cut a shape as shown and unfold it. The resulting shape has two axes of symmetry.

EXERCISE 11B   How many axes of symmetry are there in each of the following shapes?

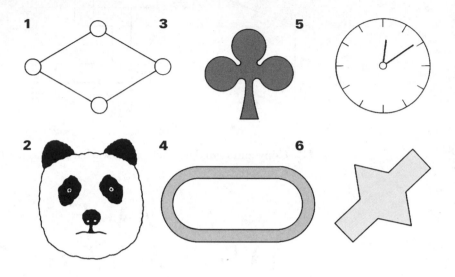

Copy the following drawings on squared paper and complete them so that the two broken lines are the two axes of symmetry.

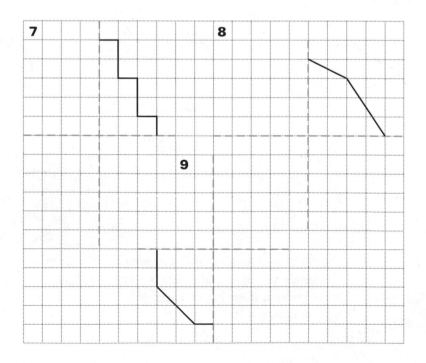

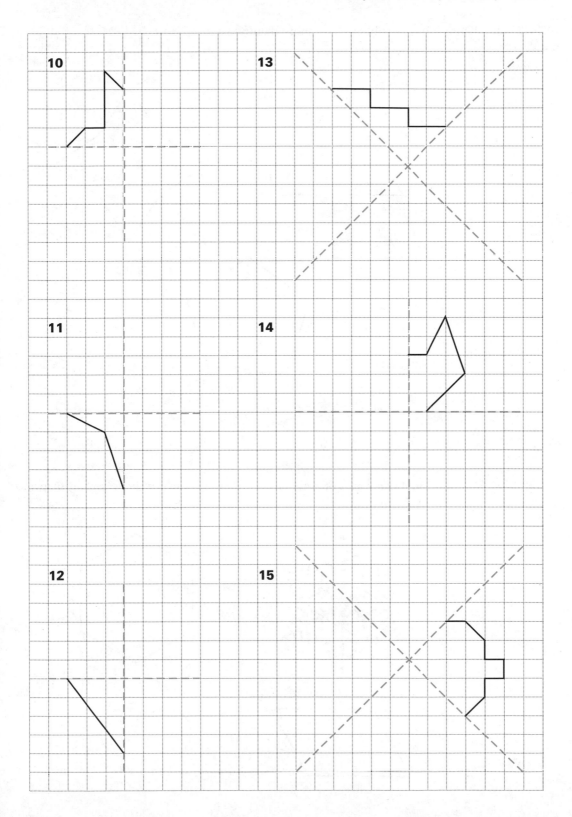

10

13

11

14

12

15

THREE OR MORE AXES OF SYMMETRY

It is possible to have more than two axes of symmetry.

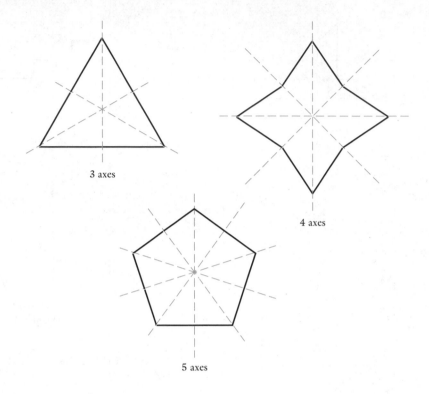

3 axes

4 axes

5 axes

EXERCISE 11C

How many axes of symmetry are there for each of the following shapes?

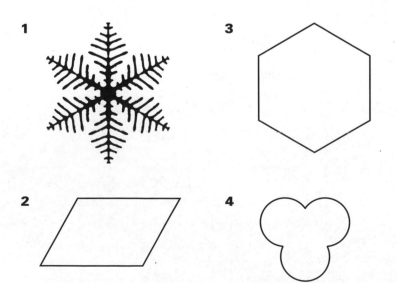

1

2

3

4

5 Fold a square piece of paper twice then fold it a third time along the broken line. Cut a shape, simple or complicated, and unfold the paper. How many axes of symmetry does it have?

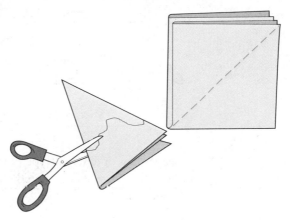

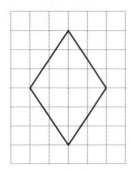

6

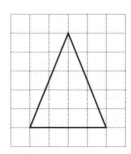

Copy the triangle on squared paper and mark in the axis of symmetry.

A triangle with an axis of symmetry is called an *isosceles triangle*.

7

Copy the quadrilateral on squared paper and mark in the two axes of symmetry.

This quadrilateral (which has four equal sides) is called a *rhombus*.

8 Trace the triangle. Draw its axes of symmetry. Measure its three sides.

This triangle is called an *equilateral triangle*.

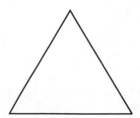

ROTATIONAL
SYMMETRY

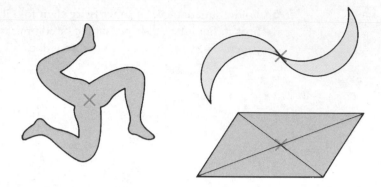

These shapes have a different type of symmetry. They cannot be folded
in half but can be turned or rotated about a centre point (marked
with ✕) and still look the same.

EXERCISE 11D

1 Lay a piece of tracing paper over any one of the shapes above, trace it
and turn it about the cross until it fits over the shape again.

Which of the following shapes have rotational symmetry?

2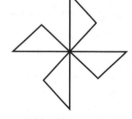

4

6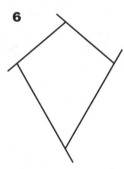

3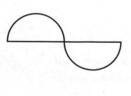

5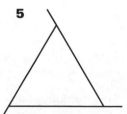

7

Some shapes have both line symmetry and rotational symmetry:

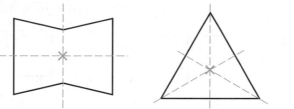

8 Sketch the capital letters of the alphabet. Mark any axes of symmetry and the centre of rotation if it exists.

For instance, draw H.

9 Which of the shapes in Exercise 11C have rotational symmetry?

CONGRUENCE The diagram below was drawn on a computer. Triangle A was drawn first. The other triangles were obtained by copying A in various ways, including turning it over and round.

All the triangles are exactly the same shape and size. If they were cut out, they would fit exactly over each other.

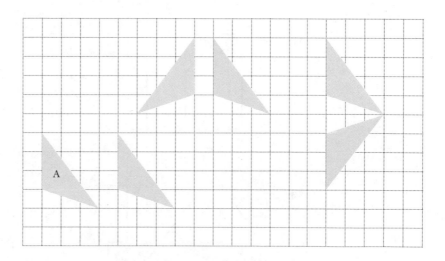

> Any two figures that are exactly the same shape and size are called *congruent* figures.

EXERCISE 11E   In each of the following questions, state whether or not the two figures are congruent. If you are not sure, trace one figure and see if the tracing will fit over the other figure. If necessary, turn the tracing over.

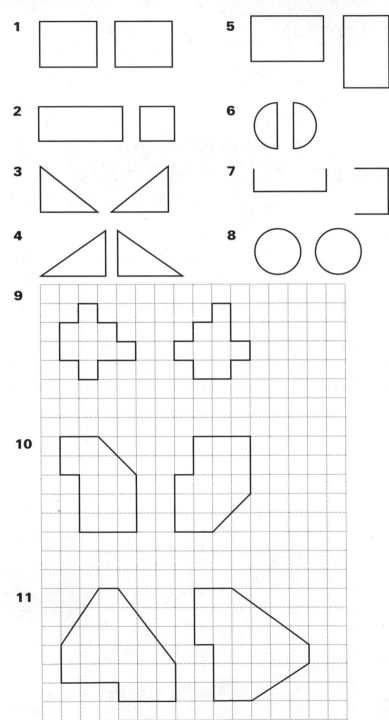

SECTIONS

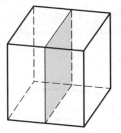

Imagine cutting straight through the middle of a cube as shown in the diagram.

The cut face, which is shaded, is called a *section* of the cube. In this case the section is a square.

When we make a straight cut through a solid, we say that the solid is cut by a plane, i.e. the section is flat.

EXERCISE 11F

1 A ball is cut by a plane (i.e. a flat surface) passing through the centre. Draw the section.

2 A cylinder is cut by a plane as shown in the diagram. Draw the section.

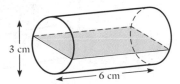

3 cm

6 cm

3 The cube is cut into two pieces by the plane through A, B, C and D.

Sketch the section, marking in any lengths and angles that you know. Are the sides of the section all the same length?

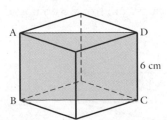

A

D

6 cm

B

C

4 Draw the shape of the section given by slicing down through the middle of each of the following solids.

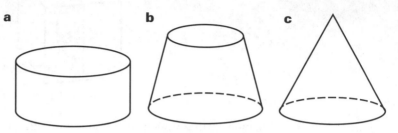

a   b   c

PLANES OF SYMMETRY

Imagine that a rectangular block is cut into two pieces by a plane.

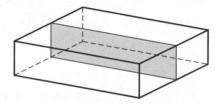

Now take one of the pieces and put the cut face against a mirror.

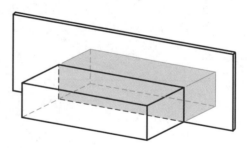

Does the piece, together with its reflection, look like the complete solid?

If it does, as happens in this case, then the cut has been made in a *plane of symmetry*.

Not all planes which cut a solid in half are planes of symmetry. The block below has *not* been cut in a plane of symmetry. If *one* half is placed against a mirror we do not see a rectangular block.

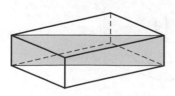
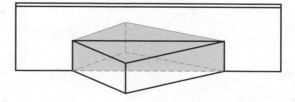

EXERCISE 11G In questions **1** to **6** state whether or not the shaded section is in a plane of symmetry of the object.

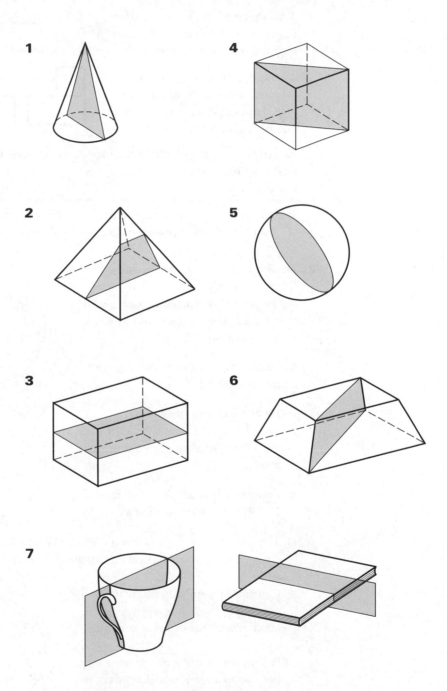

1

4

2

5

3

6

7

A cup and an exercise book each have an approximate plane of symmetry. Name other objects that have a plane of symmetry.

INVESTIGATIONS

1 In this investigation squares must meet along complete edges.

Two squares can be put together to give just one shape that has at least one line of symmetry.

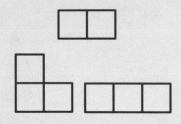

There are two ways in which three squares can be arranged to give shapes that have at least one line of symmetry.

Investigate how many different shapes can be made that have at least one line of symmetry, when 4, 5, 6, ... squares are used to make a shape.

Is there any connection between the number of squares used and the number of different shapes that satisfy the given condition?

2 Draw a clock face without hands, like the one opposite. Although there are no figures on the clock face assume that the 12 is at the top, that is, in its normal position.

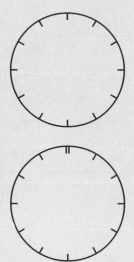

a How many axes of symmetry does this face have?

b How many axes of symmetry would it have if the position of the 12 was marked with a double line?

c Repeat parts **a** and **b** if the figures 1 to 12 are added to the face.

d The time on this clock face is 6 o'clock. What we see is symmetrical about a vertical line drawn through the centre of the clock.
Can you find any other time when the position of the hands is symmetrical about this vertical line through the centre?
Would your answer be the same if the hour hand and minute hand were the same length?
Illustrate your answers with sketches.

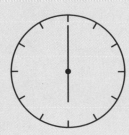

e The sketch shows that the time is 2.25.
Make a copy and mark the position of
the hour hand and the position of the
minute hand when they are reflected
in the broken line. (It is probably
better to do this in a different colour.)
Do the reflected hands give an
acceptable time?
If so, what time is it?

f

Actual time	'Reflected' time
1.55	
2.50	
3.45	
4.40	
5.35	
6.30	
7.25	
8.20	
9.15	
10.10	
11.05	

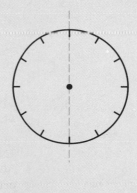

Copy this table and for each line shown

  i draw this clock face and mark in the time

  ii mark (in a different colour) the position of the hands if they
    are reflected in the broken line

  iii complete the corresponding 'reflected' time in the table.

Can you see a connection between the actual times and the
'reflected' times?

g Draw the horizontal broken line through the centre.
Mark the time 1.10.
Draw the position of the hands if
they are reflected in the broken line.
Is the 'reflected' time an acceptable
time? Justify your answer.

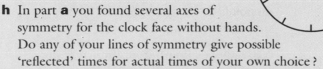

h In part **a** you found several axes of
symmetry for the clock face without hands.
Do any of your lines of symmetry give possible
'reflected' times for actual times of your own choice?

TRIANGLES AND QUADRILATERALS

12

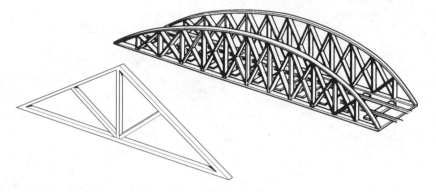

The diagrams show part of a railway bridge and a roof truss for a house.
- After these have been designed they need to be drawn accurately before they can be made.
- A modern bridge builder would produce accurate drawings using a computer but the roof truss is still simple enough to draw using pencil and paper methods.

To make accurate drawings of simple shapes you need
  a *sharp* pencil
  a ruler
  a pair of compasses
  a protractor.

USING A PAIR OF COMPASSES

We use a pair of compasses to draw a circle or an arc of a circle.

The edge of a circle is called the *circumference*.

Part of the circumference of a circle is called an *arc*.

All points on the circumference are the same distance from the centre.

A straight line drawn from the centre of a circle to the circumference is a *radius*.

A straight line drawn through the centre of a circle from edge to edge is a *diameter*.

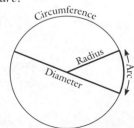

Using a pair of compasses needs practice.

Draw several circles.
Make some of them small and some large.
You should not be able to see the place
where you start and finish.

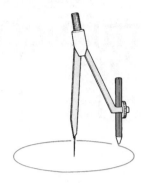

Draw a circle of radius 7 cm.
Draw a diameter of the circle and measure it.
Is it the correct length?

Now try drawing the daisy pattern below.

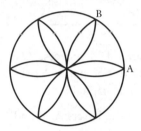

Draw a circle of radius 5 cm. Keeping the
radius the same, put the point of the
compasses at A and draw an arc to meet the
circle in two places, one of which is B. Move
the point to B and repeat. Carry on moving
the point of your compasses round the circle
until the pattern is complete.

Repeat the daisy pattern but this time draw complete circles instead of
arcs.

There are some more patterns using compasses on pages 231 and 232.

**DRAWING
STRAIGHT LINES
OF A GIVEN
LENGTH**

To draw a straight line that is 5 cm long, start by using your ruler to draw
a line that is *longer* than 5 cm.
Then mark a point on this line near one end as shown. Label it A.

A

Next set your compasses to
measure 5 cm on your ruler.

Then put the point of the
compasses on the line at A and
draw an arc to cut the line as
shown.

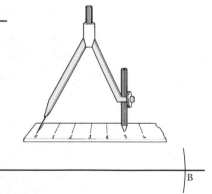

A
B

The length of line between A and B should be 5 cm. Measure it with
your ruler.

EXERCISE 12A   Draw, as accurately as you can, straight lines of the following lengths.

1 6 cm   **3** 12 cm   **5** 8.5 cm   **7** 4.5 cm

2 2 cm   **4** 9 cm   **6** 3.5 cm   **8** 6.8 cm

TRIANGLES   A triangle has three sides and three angles.

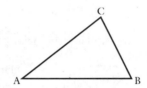

The corners of the triangle are called vertices. (One corner is called a vertex.) So that we can refer to one particular side, or to one particular angle, we label the vertices using capital letters.

In the diagram above we used the letters A, B and C so we can now talk about 'the triangle ABC' or '△ABC'.

The side between A and B is called 'the side AB' or AB.

The side between A and C is called 'the side AC' or AC.

The side between B and C is called 'the side BC' or BC.

The angle at the corner A is called 'angle A' or $\hat{A}$ for short.

We can also describe the angle at A using the three letters on the arms that enclose it, with A in the middle, i.e. $C\hat{A}B$ or $B\hat{A}C$.

EXERCISE 12B   **1** Write down the name of the side which is 4 cm long.

Write down the name of the side which is 2 cm long.

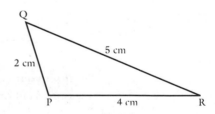

2 Write down the name of

a the side which is 2.5 cm long

b the side which is 2 cm long

c the angle which is 70°

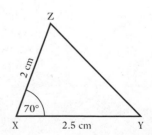

In the following questions, draw a freehand copy of the triangle and mark the given measurements on your drawing.

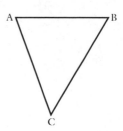

3 In $\triangle ABC$, $AB = 4\,cm$, $\widehat{B} = 60°$, $\widehat{C} = 50°$.

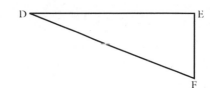

4 In $\triangle DEF$, $\widehat{E} = 90°$, $\widehat{F} = 70°$, $EF = 3\,cm$.

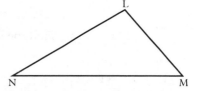

5 In $\triangle LMN$, $\widehat{L} = 100°$, $\widehat{N} = 30°$, $NL = 2.5\,cm$.

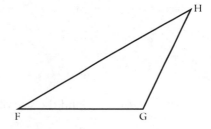

6 In $\triangle FGH$, $FG = 3.5\,cm$, $GH = 3\,cm$, $\widehat{H} = 35°$.

Make a freehand drawing of the following triangles. Label each one and mark the measurements given.

7 $\triangle ABC$ in which $AB = 10\,cm$, $BC = 8\,cm$ and $\widehat{B} = 60°$

8 $\triangle PQR$ in which $\widehat{P} = 90°$, $\widehat{Q} = 30°$ and $PQ = 6\,cm$

9 $\triangle DEF$ in which $DE = 8\,cm$, $\widehat{D} = 50°$ and $DF = 6\,cm$

10 $\triangle XYZ$ in which $XY = 10\,cm$, $\widehat{X} = 30°$ and $\widehat{Y} = 80°$

ANGLES OF A
TRIANGLE

Draw a large triangle of any shape. Use a straight edge to draw the sides. Measure each angle in this triangle, turning your page to a convenient position when necessary. Add up the sizes of the three angles.

Draw another triangle of a different shape. Again measure each angle and then add up their sizes.

Now try this: on a piece of paper draw a triangle of any shape and cut it out. Next tear off each corner and place the three corners together.

They should look like this:

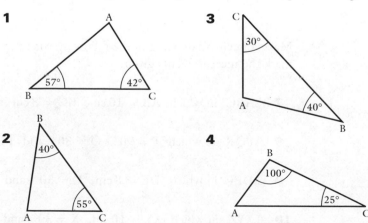

The three angles of a triangle add up to 180°.

EXERCISE 12C

Find the size of angle A.

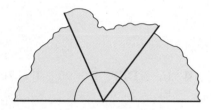

$\widehat{A} + 57° + 42° = 180°$ (angles of $\triangle$ add up to 180°)

$\therefore \qquad \widehat{A} = 180° - 99°$

$\qquad = 81°$

$\begin{array}{r} 57 \\ +42 \\ \hline 99 \\ \hline \end{array}$

Find the size of angle A (an angle marked with a square is a right angle).

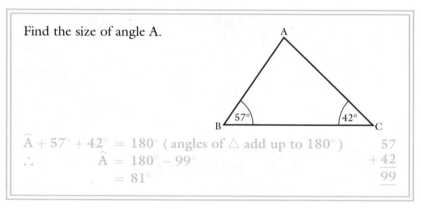

1

2

3

4

5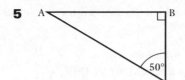

10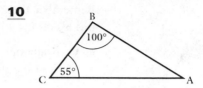

6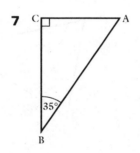

11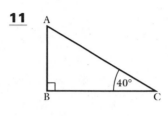

7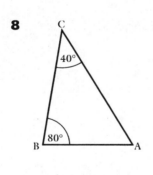

12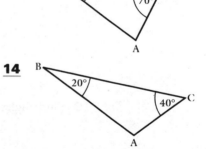

8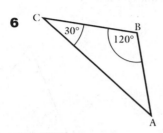

13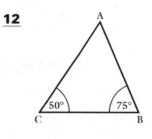

14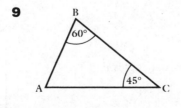

9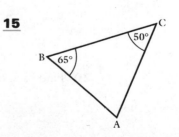

15

Remember these facts which are needed for the remaining questions in this exercise:

- Vertically opposite angles are equal.
- Angles on a straight line add up to 180°.

In each question make a rough copy of the diagram and mark the sizes of the angles that you are asked to find.

16 Find angles d and f.

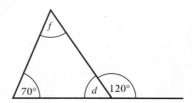

17 Find angles s and t.

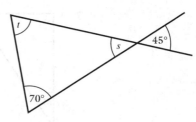

18 Find each of the equal angles x.

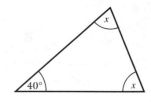

19 Find angles p and q.

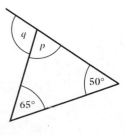

20 Find each of the equal angles g.

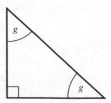

21 Find each of the equal angles *x*.

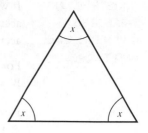

22 Find angles *s* and *t*.

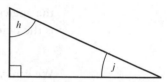

23 Angle *h* is twice angle *j*.
Find angles *h* and *j*.

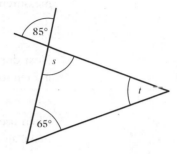

24 Find each of the equal angles *q*
and angle *p*.

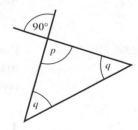

25

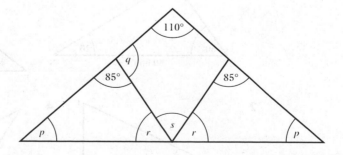

This section of a roof is symmetrical about the vertical line through
its highest point. Find the angles marked *p*, *q*, *r* and *s*.

CONSTRUCTING TRIANGLES GIVEN ONE SIDE AND TWO ANGLES

If we are given enough information about a triangle we can make an accurate drawing of that triangle. The mathematical word for 'make an accurate drawing of' is 'construct'.

For example: construct $\triangle ABC$ in which $AB = 7\,cm$, $\widehat{A} = 30°$ and $\widehat{B} = 40°$.

First make a rough sketch of $\triangle ABC$ and put all the given measurements in your sketch.

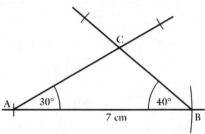

Next draw the line AB making it 7 cm long. Label the ends.

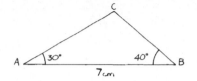

Then use your protractor to make an angle of 30° at A.

Next make an angle of 40° at B. If necessary extend the arms of the angles until they cross; this is the point C.

We can calculate $A\widehat{C}B$ because all three angles add up to $180°$ so $A\widehat{C}B = 110°$. Now as a check we can measure $A\widehat{C}B$ in our construction.

EXERCISE 12D

Construct the following triangles. Calculate the third angle in each triangle and then measure this angle to check the accuracy of your construction.

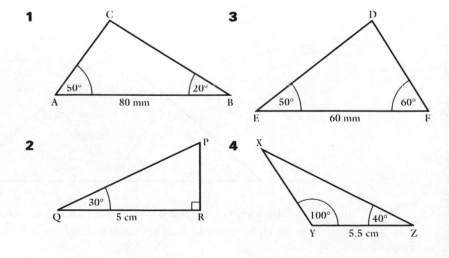

5   △UVW in which $\widehat{V} = 35°$, VW $= 5.5$ cm, $\widehat{W} = 75°$

6   △FGH in which $\widehat{F} = 55°$, $\widehat{G} = 70°$, FG $= 4.5$ cm

7   △KLM in which KM $= 10$ cm, $\widehat{K} = 45°$, $\widehat{M} = 45°$

CONSTRUCTING TRIANGLES GIVEN TWO SIDES AND THE ANGLE BETWEEN THE TWO SIDES

To construct △PQR in which PQ $= 4.5$ cm, PR $= 5.5$ cm and $\widehat{P} = 35°$, first draw a rough sketch of △PQR and put in all the measurements that you are given.

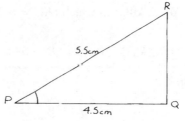

Draw one of the sides whose length you know; we will draw PQ.

Now using your protractor make an angle of 35° at P. Make the arm of the angle quite long.

Next use your compasses to measure the length of PR on your ruler.

Then with the point of your compasses at P, draw an arc to cut the arm of the angle. This is the point R.

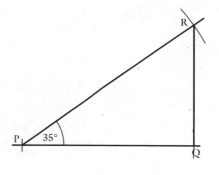

Now join R and Q.

EXERCISE 12E

Construct each of the following triangles and measure the third side.

1

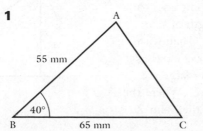

2

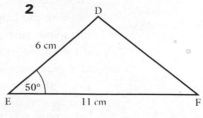

3

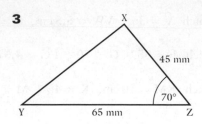

4

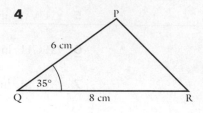

5 △HJK in which HK = 4.2 cm, $\hat{H}$ = 45°, HJ = 5.3 cm

6 △ABC in which AC = 6.3 cm, $\hat{C}$ = 48°, CB = 5.1 cm

7 △XYZ in which $\hat{Y}$ = 65°, XY = 3.8 cm, YZ = 4.2 cm

CONSTRUCTING TRIANGLES GIVEN THE LENGTHS OF THE THREE SIDES

To construct △XYZ in which XY = 5.5 cm, XZ = 3.5 cm and YZ = 6.5 cm, first draw a rough sketch of the triangle and put in all the given measurements.

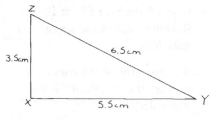

Next draw one side; we will draw XY.

Then with your compasses measure the length of XZ from your ruler. With the point of your compasses at X draw a wide arc.

Next use your compasses to measure the length of YZ from your ruler. Then with the point of your compasses at Y draw another large arc to cut the first arc. Where the two arcs cross is the point Z. Join ZX and ZY.

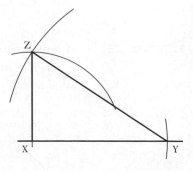

EXERCISE 12F   Construct the following triangles.

1

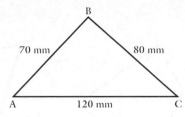

3

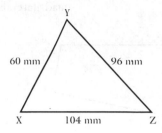

2

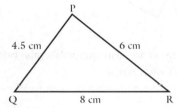

4

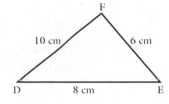

5 △ABC in which AB = 7.3 cm, BC = 6.1 cm, AC = 4.7 cm

6 △DEF in which DE = 10.4 cm, EF = 7.4 cm, DF = 8.2 cm

7 △PQR in which PQ = 8.8 cm, QR = 6.6 cm, PR = 11 cm

EXERCISE 12G   Construct the following triangles. Remember to draw a rough diagram of the triangle first and then decide which method you need to use.

1 △ABC in which AB = 7 cm, $\widehat{A} = 30°$, $\widehat{B} = 50°$

2 △PQR in which PQ = 50 mm, QR = 40 mm, RP = 70 mm

3 △BCD in which $\widehat{B} = 60°$, BC = 5 cm, BD = 4 cm

4 △WXY in which WX = 5 cm, XY = 6 cm, $\widehat{X} = 90°$

5 △KLM in which KL = 64 mm, LM = 82 mm, KM = 126 mm

6 △ABC in which $\widehat{A} = 45°$, AC = 8 cm, $\widehat{C} = 110°$

7 △CDE in which CD = DE = 60 mm, $\widehat{D} = 60°$

8 Try to construct a triangle ABC in which $\widehat{A} = 30°$, AB = 5 cm, BC = 3 cm.

9 Construct two triangles which fit the following measurements:
△PQR in which $\widehat{P} = 60°$, PQ = 6 cm, QR = 5.5 cm.

10 Construct △ABC in which $\widehat{A} = 120°$, AB = 4 cm, BC = 6 cm. Can you construct more than one triangle that fits these measurements?

QUADRILATERALS A quadrilateral has four sides. These shapes are examples of quadrilaterals:

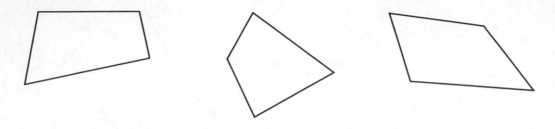

The following diagrams are also quadrilaterals, but each one is a 'special' quadrilateral with its own name:

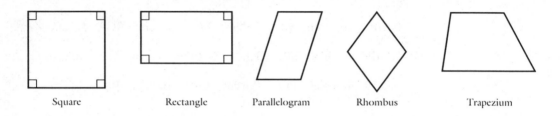

| Square | Rectangle | Parallelogram | Rhombus | Trapezium |

Draw yourself a large quadrilateral, but do not make it one of the special cases. Measure each angle and then add up the sizes of the four angles.

Do this again with another three quadrilaterals.

Now try this: on a piece of paper draw a quadrilateral. Tear off each corner and place the vertices together. It should look like this:

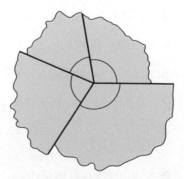

The sum of the four angles of a quadrilateral is 360°.

This is true of any quadrilateral whatever its shape or size.

EXERCISE 12H

Make a rough copy of the following diagrams and mark on your diagram the sizes of the required angles. You can also write in the sizes of any other angles that you may need to find.

In questions **1** to **10** find the size of the angle marked *d*.

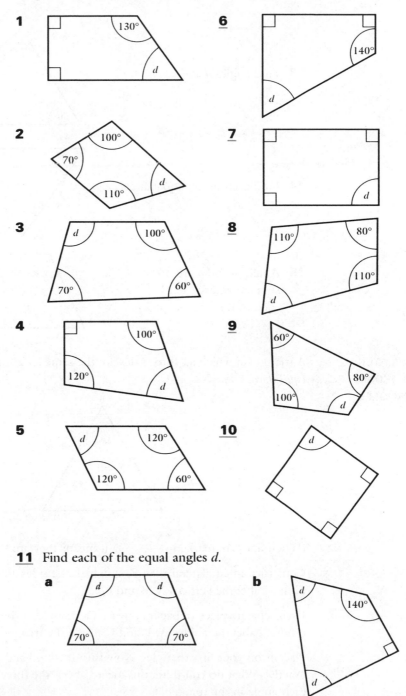

1

6

2

7

3

8

4

9

5

10

11 Find each of the equal angles *d*.

a

b

12 Angle *e* is twice angle *d*.
Find angles *d* and *e*.

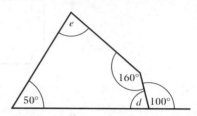

13 Find angles *d* and *e*.

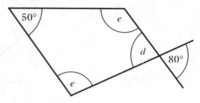

14 Find each of the equal angles *e*.

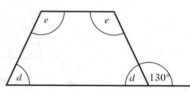

15 Angles *d* and *e* are
supplementary. Find
each of the equal angles *e*.

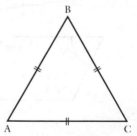

**EQUILATERAL
AND ISOSCELES
TRIANGLES**

A triangle in which all three sides are the same length is called an
equilateral triangle.

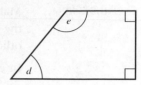

(When lines are marked ——╫——, it means they are the same length.)

Construct an equilateral triangle in which the sides are each of length
6 cm. Label the vertices A, B and C.

On a separate piece of paper construct a triangle of the same size and cut
it out. Label the angles A, B and C inside the triangle.

Place it on your first triangle. Now turn it round and it should still fit
exactly. What do you think this means about the three angles? Measure
each angle in the triangle.

In an equilateral triangle all three
sides are the same length and
each of the three angles is 60 °.

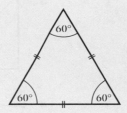

A triangle in which two sides are equal is called an *isosceles triangle*.

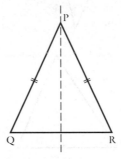

On a piece of paper construct an isosceles triangle PQR in which
$PQ = 8\,cm$, $PR = 8\,cm$ and $\widehat{P} = 80°$. Cut it out and fold the triangle
through P so that the corners at Q and R meet. You should find
that $\widehat{Q} = \widehat{R}$. (The fold line is a line of symmetry.)

In an isosceles triangle two sides
are equal and the two angles at the
base of the equal sides are equal.

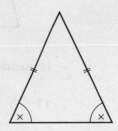

EXERCISE 12I

In questions **1** to **10** make a rough sketch of the triangle and mark
angles that are equal.

1

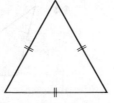

2

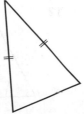

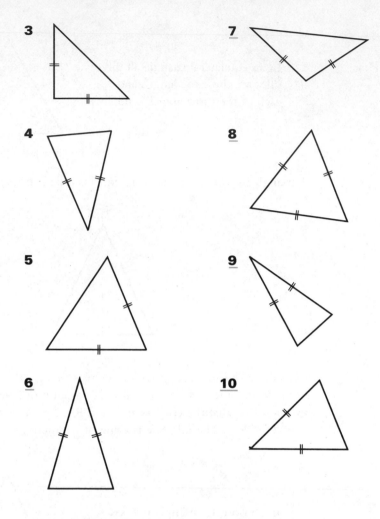

In questions **11** to **22** find angle *d*. Give reasons for your working.

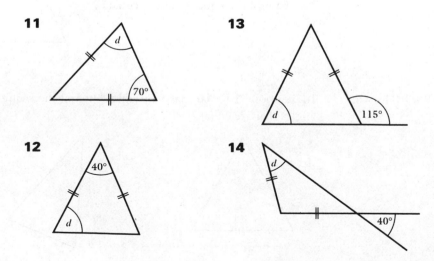

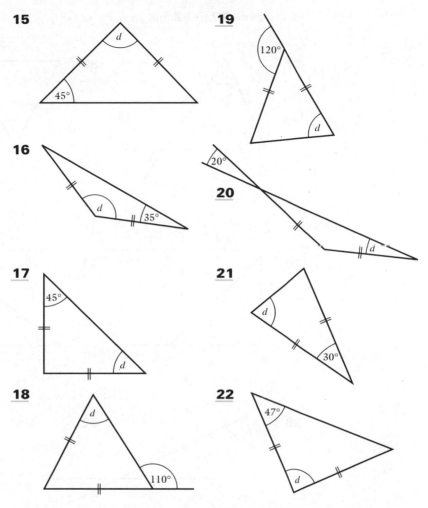

15

16

17

18

19

20

21

22

In questions **23** to **26** make a rough sketch and mark the equal sides.

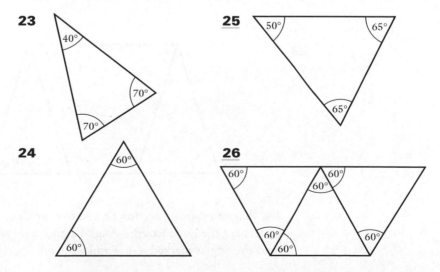

23

24

25

26

In questions **27** to **32** find angles *d* and *e*.

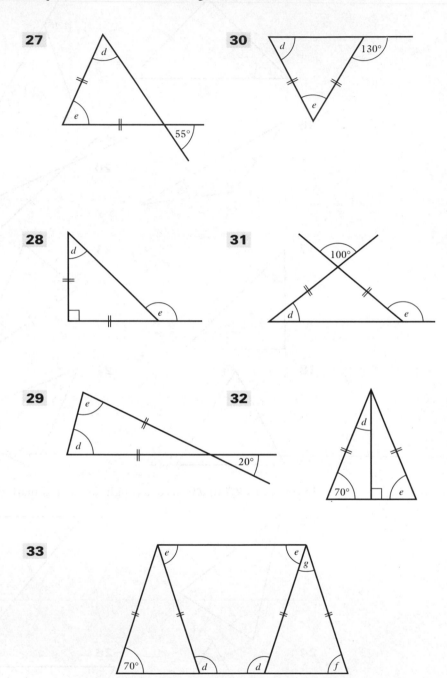

27

30

28

31

29

32

33

The diagram shows a section from a road bridge. The four inclined girders are the same length. Angles *d* and *e* are supplementary. Find the angles marked *d*, *e*, *f* and *g*.

MIXED EXERCISE

EXERCISE 12J

1 Find the size of the angle marked x.

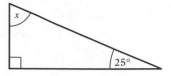

2

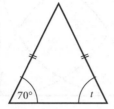

Find the size of the angle marked t.

3 Find the size of the angle marked y.

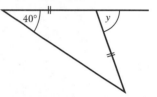

4 Construct $\triangle$ABC in which AB = 6 cm, BC = 4 cm and $\widehat{B} = 40°$. Measure AC.

5 Construct $\triangle$ABC in which BC = 10 cm, AB = 6 cm, AC = 8 cm. Measure $\widehat{A}$.

PRACTICAL WORK

The patterns below are made using a pair of compasses. Try copying them. Some instructions are given which should help.

1

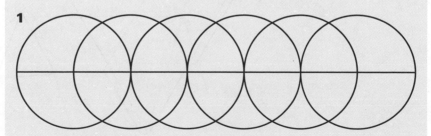

Draw a straight line. Open your compasses to a radius of 3 cm and draw a circle with its centre on the line. Move the point of the compasses 3 cm along the line and draw another circle. Repeat as often as you can.

2   Draw a square of side 4 cm. Open your compasses to a radius of 4 cm and with the point on one corner of the square draw an arc across the square. Repeat on the other three corners.

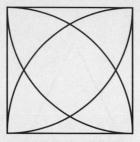

Try the same pattern, but leave out the sides of the square; just mark the corners. A block of four of these looks good.

3   On a piece of paper construct an equilateral triangle of side 4 cm. Construct an equilateral triangle, again of side 4 cm, on each of the three sides of the first triangle. Add tabs as shown.

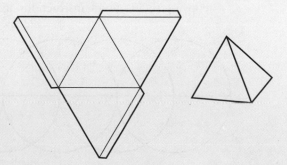

Cut out the complete diagram. Fold the outer triangles up so that the corners meet. Stick the edges together using the tabs. You have made a tetrahedron. (These make good Christmas tree decorations if made out of foil-covered paper.)

4

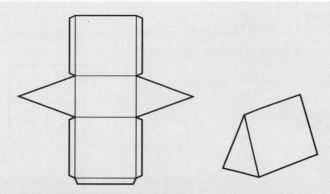

On a sheet of paper draw a rectangle measuring 6 cm by 4 cm.
On each of the shorter sides construct an isosceles triangle with
sloping sides of length 6 cm. On each of the longer sides of the
rectangle draw a square of side 6 cm. Draw tabs on the sides as shown.
Cut out the complete figure. Fold up the squares so that their outer
edges meet and fold up the triangles to make the ends.
Use the tabs to stick the edges together.

You have made a triangular prism.

PUZZLES

1 This cross can be divided into
4 identical pieces in at least
3 different ways.
Draw diagrams to show how
this is possible.

2

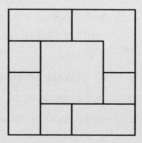

Eight serviettes are placed flat but overlapping on a table and give
the outlines shown in the diagram. In which order must they be
removed if the top one is always taken off next?

1 Five different triangles can be drawn on a 2 × 2 grid.

Two of them are shown. Sketch the other three.
Remember that is the same as

Investigate the number of different triangles that can be drawn on a 3 × 3 grid.

Extend your investigation to a 4 × 4 grid and then to a 5 × 5 grid. Can you find a connection between the number of different triangles that can be drawn and the length of the side of the grid?

2 The diagram shows three rectangles which have been formed by shading some of the squares in this 4 × 4 square. How many different rectangles can you find?
Remember that a square is a special kind of rectangle.

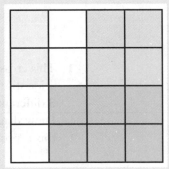

Investigate the number of rectangles there are in other squares. Can you find a connection, expressed in words, between the length of a side of a square and the number of different rectangles that can be found in it?

Hints

a Start with the simplest possible square.

b Increase the lengths of the sides of the square 1 unit at a time.

c Try to count up the number of different rectangles in an orderly way.

d When you think you have found a connection, test it to see if it gives the correct answers for squares you have already investigated and for a square you have not yet investigated.

PROBABILITY

CHANCE

Will you win a prize on the National Lottery?
If you are under 16, the answer should be NO!
For someone legally buying just one ticket, the answer is VERY UNLIKELY!

Questions such as

- Will it rain on my way home from school tonight?

- Will Manchester United win their next home game?

- Will the next bus to arrive at this stop take me to school?

- If I toss this coin will it land heads up?

have answers that can range from NO, through MAYBE, to YES.

The answer is NO when the event described is impossible, YES when the event is certain and MAYBE when there is chance involved.

If we know something about the event described, we may describe the chance that it will happen using words such as 'improbable', 'evens', 'fairly likely'.
For example, we know that a coin can land either head up or tail up and, unless the coin is faulty, each possibility is equally likely, so the fourth question might be answered 'It is as likely as not.'
To answer the third question, we need 'local' knowledge, such as the number of routes that pass the school, how often the buses run, and so on.

EXERCISE 13A

Discuss each event with the other members of a group or class.
Decide first if each event is IMPOSSIBLE, MAY HAPPEN, or is CERTAIN.
For the events that may happen, use your own words, where you can, to describe the chances that they will happen. Where you cannot, discuss what information is needed to be in a position to quantify the chance.

1 A buttered slice of bread will land butter side down when dropped.

2 You will go to the cinema next Saturday.

3 If you throw an ordinary six-sided dice it will score 7.

4 The next sweet in a tube of fruit pastilles is red.

5 It will be dark at midnight tonight.

6 The next vehicle to pass the school gate will be a 46-seater coach.

OUTCOMES OF EXPERIMENTS

Before we can start to quantify chance, we need to know all the possibilities.

If you throw an ordinary dice there are six possible scores that you can get. These are 1, 2, 3, 4, 5, and 6.

The act of throwing the dice is called an *experiment*.

The score that you get is called an *outcome* or an *event*.

The list 1, 2, 3, 4, 5, 6, is called the set of *all possible outcomes*.

EXERCISE 13B

How many possible outcomes are there for the following experiments? In each case, write down the list of all possible outcomes.

1 Tossing a 10 p coin. (Assume that it lands flat.)

2 Taking one disc from a bag containing 1 red, 1 blue and 1 yellow disc.

3 Choosing one number from the first ten positive integers. (An integer is a whole number.)

4 Taking one crayon from a box containing 1 red, 1 yellow, 1 blue, 1 brown, 1 black and 1 green crayon.

5 Taking one item from a bag containing 1 packet of chewing gum, 1 packet of boiled sweets and 1 bar of chocolate.

6 Taking one coin from a bag containing one 1 p coin, one 10 p coin, one 20 p coin and one 50 p coin.

7 Choosing one card from part of a pile of ordinary playing cards containing just the suit of clubs.

8 Choosing one letter from the vowels of the alphabet.

9 Choosing one number from the first 5 prime numbers.

10 Choosing an even number from the first 20 positive integers.

PROBABILITY

If you throw an ordinary dice, what are the chances of getting a four? If you throw it fairly, it is reasonable to assume that you are as likely to throw any one score as any other, i.e. all outcomes are equally likely. As throwing a four is only 1 of the 6 equally likely outcomes you have a 1 in 6 chance of throwing a four.

'Odds' is another word in everyday language that is used to describe chances. In mathematical language we use the word '*probability*' to describe chances.

We say that the probability of throwing a four is $\frac{1}{6}$.
This can be written more briefly as

$$P(\text{throwing a four}) = \frac{1}{6}$$

We will now define exactly what we mean by 'the probability that something happens'.

If A stands for a particular event, the probability of A happening is written $P(A)$ where

$$P(A) = \frac{\text{the number of ways in which } A \text{ can occur}}{\text{the total number of equally likely outcomes}}$$

We can use this definition to work out, for example, the probability that if one card is drawn at random from a full pack of ordinary playing cards, it is the ace of spades.
(The phrase '*at random*' means that any one card is as likely to be picked as any other.)

There are 52 cards in a full pack, so there are 52 equally likely outcomes. There is only one ace of spades, so there is only one way of drawing that card, i.e.

$$P(\text{ace of spades}) = \frac{1}{52}$$

EXERCISE 13C   In the following questions, assume that all possible outcomes are equally likely.

1 One letter is chosen at random from the letters in the word SALE. What is the probability that it is A?

2 What is the probability that a red pencil is chosen from a box containing 10 different coloured pencils?

3 What is the probability of choosing a prime number from the numbers 6, 7, 8, 9, 10?

4 What is the probability of picking the most expensive car from a range of six new cars in a showroom?

5 What is the probability of choosing an integer that is exactly divisible by 5 from the list 6, 7, 8, 9, 10, 11, 12?

6 In a raffle 200 tickets are sold.
If you have bought one ticket, what is the probability that you will win first prize?

7 One card is chosen at random from a pack of 52 ordinary playing cards.
What is the probability that it is the ace of hearts?

8 What is the probability of choosing the colour blue from the colours of the rainbow?

9 A whole number is chosen from the first 15 positive whole numbers. What is the probability that it is exactly divisible both by 3 and by 4?

EXPERIMENTS WHERE AN EVENT CAN HAPPEN MORE THAN ONCE

If a card is picked at random from an ordinary pack of 52 playing cards, what is the probability that it is a 'five'?

There are 4 fives in the pack, the five of spades, the five of hearts, the five of diamonds and the five of clubs.

So there are 4 ways in which a five can be picked.
Altogether there are 52 cards that are equally likely to be picked,

therefore $\qquad$ P (picking a five) $= \dfrac{4}{52} = \dfrac{1}{13}$

Now consider a bag containing 3 white discs and 2 black discs.

If one disc is taken from the bag it can be black or white. But these are not equally likely events: there are three ways of choosing a white disc and two ways of choosing a black disc, so

$$P(\text{ choosing a white disc }) = \frac{3}{5}$$

and

$$P(\text{ choosing a black disc }) = \frac{2}{5}$$

EXERCISE 13D

A letter is chosen at random from the letters of the word DIFFICULT.
What is the probability that the letter I will be chosen?

$$P(\text{ choosing I }) = \frac{2}{9}$$

There are 2 ways of choosing the letter I and there are 9 letters in DIFFICULT, so I can be chosen in 2 out of 9 equally likely choices.

1 How many ways are there of choosing an even number from the first 10 positive whole numbers?

2 A prime number is picked at random from the list
4, 5, 6, 7, 8, 9, 10, 11.
How many ways are there of doing this?

3 A card is taken at random from an ordinary pack of 52 playing cards. How many ways are there of taking a black card?

4 An ordinary six-sided dice is thrown.
How many ways are there of getting a score that is greater than 4?

5 A lucky dip contains 50 boxes, only 10 of which contain a prize, the rest being empty.
What is the probability of choosing a box that contains a prize?

6 A number is chosen at random from the first 10 positive integers.
What is the probability that it is

a an even number   **c** a prime number

b an odd number   **d** exactly divisible by 3?

7 One card is drawn at random from an ordinary pack of 52 playing cards.
What is the probability that it is

a an ace   **c** a heart

b a red card   **d** a picture card (include the aces)?

8 One letter is chosen at random from the word INNINGS.
What is the probability that it is

a the letter N

b the letter I

c a vowel

d one of the first five letters of the alphabet?

9 An ordinary six-sided dice is thrown.
What is the probability that the score is

a greater than 3   **b** at least 5   **c** less than 3?

10 A book of 150 pages has a picture on each of 20 pages.
If one page is chosen at random, what is the probability that it has a picture on it?

11 One counter is picked at random from a bag containing 15 red counters, 5 white counters and 5 yellow counters.
What is the probability that the counter removed is

a red **b** yellow **c** not red?

12 If you bought 10 raffle tickets and a total of 400 were sold, what is the probability that you win first prize?

13 A roulette wheel is spun.
What is the probability that when it stops it will be pointing to

a an even number

b an odd number

c a number less than 10 excluding zero?

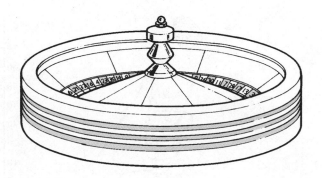

(The numbers on a roulette wheel go from 0 to 35, and zero is neither an even number nor an odd number.)

14 One letter is chosen at random from the letters of the alphabet.
What is the probability that it is a consonant?

15 A number is chosen at random from the set of two-digit numbers (i.e. the numbers from 10 to 99).
What is the probability that it is exactly divisible both by 3 and by 4?

16 A bag of sweets contains 4 caramels, 3 fruit centres, 5 mints and 4 cola bottles.
If one sweet is taken out, what is the probability that it is

a a mint **b** a caramel **c** not a fruit centre?

FINDING PROBABILITY FROM RELATIVE FREQUENCY

We have assumed that if you toss a coin it is equally likely to land head up or tail up so that $P(\text{a head}) = \frac{1}{2}$.
Coins like this are called 'fair' or 'unbiased'.

Most coins are likely to be unbiased but it is not necessarily true of all coins. A particular coin may be slightly bent or even deliberately biased so that there is not an equal chance of getting a head or a tail.

The only way to find out if a particular coin is unbiased is to collect some information about how it behaves when it is tossed. We can do this by tossing it several times and recording the results.

Tossing a bent coin gave these results.

 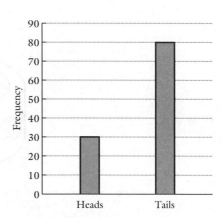

For this coin we have 30 heads out of 110 tosses.
We say that the *relative frequency* of heads with this coin is $\frac{30}{110}$.

We can use the relative frequency of heads as an estimate for the probability of a head with this coin, i.e.

$$P(\text{a head}) \approx \frac{\text{number of heads}}{\text{total number of tosses}} = \frac{30}{110} = \frac{3}{11}$$

The approximation gets more reliable as the number of tosses gets larger, but on this evidence it looks as though this coin is more likely to give a tail than a head.

Work with a partner or collect information from the whole class.
Use an ordinary six-sided dice.

1 a Throw the dice 25 times and keep a tally of the number of sixes.
Use your results to find the relative frequency of sixes.

b Now throw the dice another 25 times and add the results to the
last set. Use these to find again the relative frequency of sixes.
Now do another 25 throws and add the results to the last two
sets to find another value for the relative frequency.

c Carry on doing this in groups of 25 throws until you have done
200 throws altogether.
You know that the probability of getting a six is $\frac{1}{6}$. Now look
at the sequence of results obtained from your experiment.
What do you notice? (It is easier to compare your results if you
use your calculator to change the fractions into decimals to
2 d.p.)

2 a A dice is to be thrown 60 times and the numbers that appear
are to be recorded. Roughly how many times do you expect
each of the numbers 1 to 6 to appear?

b Now throw a dice 60 times.
Record the results on a tally chart and draw a bar chart.
Has it come out as you expected?

c Combine your information with that of several other people, so
that you have the results of, say, 180 or 240 throws.
Draw a bar chart. Comment on its shape.

d Throw the dice 10 times and record the numbers.
Would it make sense to make a judgement about the fairness of
this dice using the relative frequencies of each score from just
these 10 results?

e Throw the dice again 10 times.
Has the same set of numbers been thrown as in part **d**?

f Imagine that the dice is thrown 10 more times.
Can you rely on getting the same numbers again as in parts
d and **e**? What extreme case might you get?

3 This question should be done as a class exercise.

A company runs the following promotional offer with tubes of sweets.

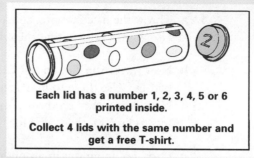

Each lid has a number 1, 2, 3, 4, 5 or 6 printed inside.

Collect 4 lids with the same number and get a free T-shirt.

What are the chances of getting four numbers the same if you just buy four tubes?

What is the most likely number of tubes that you need to buy to collect enough lids?

a We could answer the questions by collecting evidence, i.e. buying tubes of these sweets until we have four lids with the same number on them, and repeating this until we think we have reliable results.

What are the disadvantages of this?

b We can avoid the disadvantages of having to buy tubes of these sweets by simulating the situation as follows:

First assume that any one of the numbers 1 to 6 is equally likely to be inside a lid.

Now throw a dice to simulate the number we would get if we bought a tube, and carry on throwing it until you get four numbers the same. We will need several tally charts to keep a record of the number of throws needed on each occasion.

This shows the start of the simulation.

Score	Tally
1	/
2	///
3	/
4	//
5	////
6	/
Total	12

Number of throws needed to get 4 numbers the same	4	5	6	7	8	9	10	11	12	13	14	15	16	...
Tally									/					

4 Make two three-sided spinners using card and a pencil.

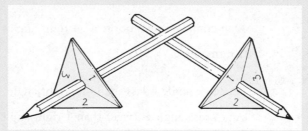

a Write down the list of possible outcomes for the sum of the scores when both spinners are spun.

b Estimate the probability of getting a total score of 3 when both spinners are used.

Comment on how reliable your estimate is.

Remember to show all your working and explain your reasoning.

SUMMARY 3

ANGLES

One complete revolution = 4 right angles = 360°.

1 right angle = 90°.

An acute angle is less than one right angle.

An obtuse angle is larger than 1 right angle but less than 2 right angles.

A reflex angle is larger than two right angles.

Vertically opposite angles are equal.

Angles on a straight line add up to 180°.

Two angles that add up to 180° are called supplementary angles.

Angles at a point add up to 360°.

SYMMETRY

A figure has line symmetry when it can be folded along the line so that one half fits exactly over the other half.
A figure has rotational symmetry when it can be turned through less than 360° about a point so that it looks the same.

CONGRUENCE

Two figures are congruent when they are exactly the same shape and size, even if one is turned over compared with the other.

TRIANGLES

The three angles in a triangle add up to 180°.

An equilateral triangle has all three sides equal and each angle is 60°.

An isosceles triangle has two sides equal and the angles at the base of these sides are equal.

QUADRILATERALS

A quadrilateral has four sides.
The four angles in a quadrilateral add up to 360°.

PROBABILITY The probability that an event *A* happens is given by

$$\frac{\text{the number of ways in which } A \text{ can occur}}{\text{the total number of equally likely outcomes}}$$

When we perform experiments to find out how often an event occurs, the relative frequency of the event is given by

$$\frac{\text{the number of times the event occurs}}{\text{the number of times the experiment is performed}}$$

Relative frequency is used to give an approximate value for probability.

REVISION
EXERCISE 3.1
(Chapters 10 & 11)

1 How many degrees does a hand of a clock turn through when it moves

   a from 7 to 12   **b** from 9 to 3   **c** from 2 to 5 ?

2 If you stand facing east and turn clockwise through $\frac{1}{4}$ of a turn, in which direction are you facing?

3 Which of these statements are true and which are false?

   a The difference between an obtuse angle and an acute angle is always less than 90°.

   b 190° is an example of an obtuse angle.

   c Every reflex angle is bigger than every obtuse angle.

   d The difference between a reflex angle and an obtuse angle is sometimes less than 90°.

4 Estimate the size of each of the following angles.

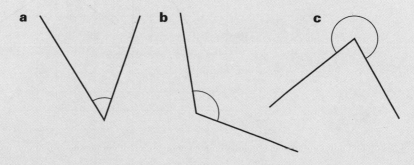

5 Find the size of each angle marked with a letter.

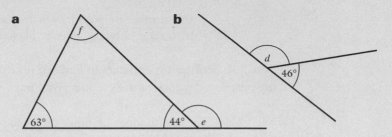

a

b

6 Find the size of each angle marked with a letter.

a

b   Angle *r* is twice angle *t*.

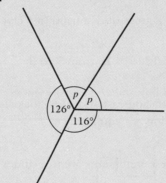

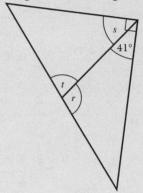

7 Copy the drawing onto squared paper and complete it so that the broken line is the axis of symmetry.

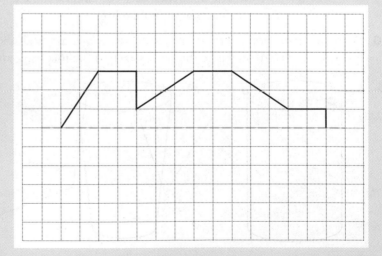

8 Copy the following drawings onto squared paper and complete them so that the broken lines are the two axes of symmetry.

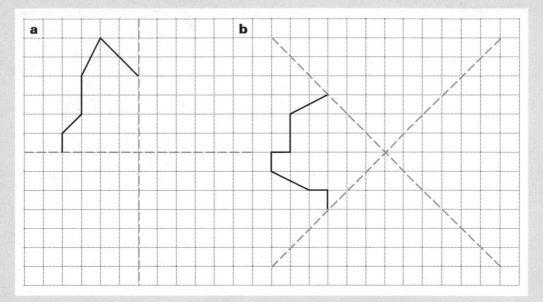

9 Which of these shapes have both line and rotational symmetry?

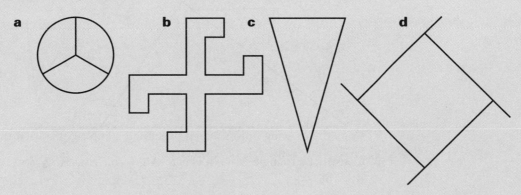

10 Draw the shape of the section given by slicing down through the middle of each of the following solids.

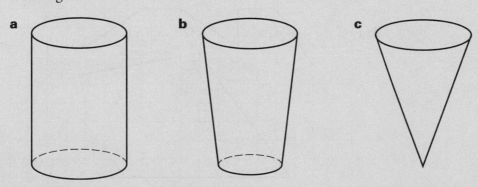

1 Find the size of the angle at A.

a

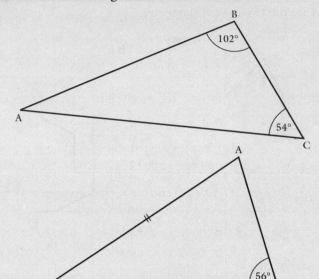

b

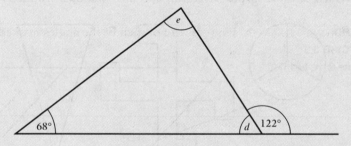

2 Find the size of the angles marked *d* and *e*.

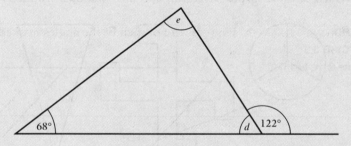

3 Construct a triangle ABC in which AB = 6.5 cm, $\widehat{A} = 55°$,
$\widehat{B} = 65°$.

4 Construct a triangle XYZ in which XY = 75 mm, YZ = 95 mm,
XZ = 80 mm.

5 Find the size of the angle marked with a letter.

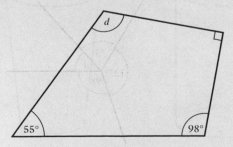

6 How many possible outcomes are there of

   a choosing one number from the first six positive whole numbers

   b choosing one number from the first ten prime numbers?

7 One letter is chosen at random from the letters in the word
               TELEVISION.
   What is the probability that the letter is

   a E   **b** a vowel   **c** a consonant?

8 What is the probability of choosing a prime number from the
   numbers 10, 11, 12, 13, 14, 15?

9 A card is chosen at random from a pack of 52 ordinary playing cards.
   What is the probability that it is

   a the ace of spades   **c** a black card

   b the 2 of spades   **d** a club?

10 Which event would you think is the more likely to happen: score 6
   when you roll a dice or draw a heart when you select a card at
   random from a pack of 52 ordinary playing cards? Give reasons for
   your answer.

**REVISION
EXERCISE 3.3
(Chapters 10 to 13)**

1 Find the size of each of the angles marked with a letter.

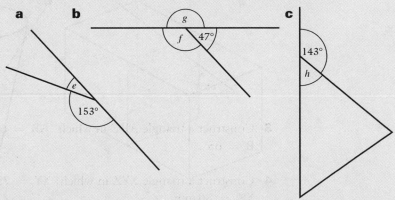

2 Find the size of each angle marked with a letter.

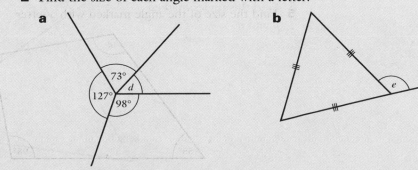

3 Find the size of each marked angle.

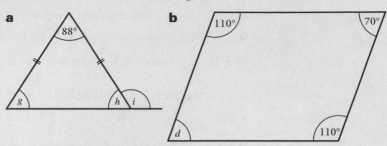

4 State whether or not these pairs of figures are congruent.

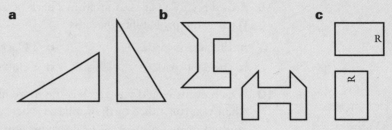

5 State whether or not the shaded section is a plane of symmetry of the object.

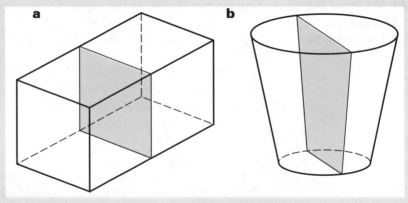

6 Find the angles marked p and q.

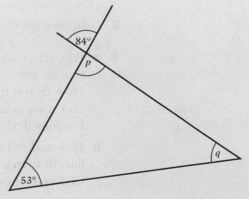

7 Construct triangle ABC in which AB = 45 mm, BC = 55 mm,
$\widehat{B}$ = 50°.

8 Construct a triangle PQR in which PQ = 9.5 cm, $\widehat{P}$ = 63°,
$\widehat{Q}$ = 72°.

9 One letter is chosen at random from the letters in the word PSST.
What is the probability that the letter is

a a vowel   **b** a consonant?

10 What is the probability of randomly choosing an integer that is
exactly divisible by 6 from the list

$$6, 8, 10, 12, 14, 16, 18, 20\,?$$

**REVISION
EXERCISE 3.4
(Chapters 1 to 13)**

1 a Find **i** $\frac{7}{12} + \frac{2}{3} + \frac{1}{6}$ **ii** $10\frac{3}{4} - 4\frac{5}{8}$

b Express 44% as **i** a decimal   **ii** a fraction in its lowest terms.

c Split 184 p equally among four sisters.

2 a Find $3 \times 4 \times 5 + 28 \div 2$

b Find $5429 \div 42$, giving the remainder if there is one.

c Use a calculator to find, correct to 2 decimal places

i 4.09×1.08 **ii** $5.92 \div 0.87$

3 John bought a loaf of bread costing 87 p and a fruit cake costing
£1.95. He paid with a £5 note.
How much change did he get?

4 Two girls set off walking together and are in step for their first pace.
One girl has a pace of 90 cm and the other has a pace of 75 cm.
How far will they walk before they are in step together again?
Give your answer in   **a** centimetres   **b** metres.

5 a Fifty cars, each 4165 mm long, are parked nose to tail in a
straight line.
How far is it from the front of the first car to the tail of the last
car? Give your answer in
i millimetres **ii** metres **iii** metres correct to the nearest metre.

b How many of these cars would be needed, parked nose to tail in a
line, to stretch a distance of 1 km?
Give your answer to the nearest whole number.

6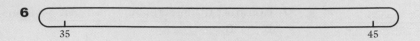

35   45

The diagram shows a glass tube that is to be scaled to show temperature between 35 °C and 45 °C. The distance between the mark for 35 and the mark for 45 is 12 cm.

a How far apart should the marks be for the temperatures 35 °C and 36 °C?

b This distance is then divided to show tenths of a degree. How far apart should these marks be?

7 Find the difference between

a 0.55 m and 136 cm   **b** 24.6 cm and 565 mm.

Give each answer in the smaller unit.

8 Which of the shapes are congruent with the coloured shape?

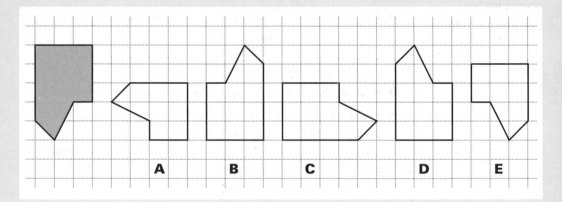

A   B   C   D   E

9 Angle *d* is twice angle *e*.
Find the angles *d* and *e*.

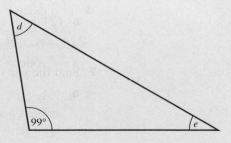

10 One letter is chosen at random from the letters in the word ASSESS. What is the probability that the letter is

a E   **b** S   **c** not S?

**REVISION
EXERCISE 3.5
(Chapters 1 to 13)**

1 A car travelling at 50 miles per hour took 4 hours to travel from Manchester to Cardiff.
How many miles did the car travel?

2 500 eggs are packed on trays, 36 to a tray.
How many trays are required?

3 The list shows the marks out of 10 obtained by a group of pupils in a test.

$$8 \quad 7 \quad 9 \quad 4 \quad 8 \quad 5 \quad 6 \quad 8 \quad 5$$
$$4 \quad 8 \quad 10 \quad 8 \quad 5 \quad 7 \quad 3 \quad 8 \quad 9$$
$$7 \quad 4 \quad 7 \quad 6 \quad 8 \quad 7 \quad 5 \quad 9 \quad 7$$

a Make a frequency table for this list.

b How many pupils were tested?

c Draw a bar chart to show the information in the frequency table.

4 a Add $1\frac{3}{4}$ to $2\frac{5}{12}$ and subtract the total from $5\frac{5}{6}$.

b To 1.08 add the difference between 7.92 and 6.34.

5  **a** How many grams of fat are there in this pot of fromage frais?

b What fraction of the weight in this pot is fat?

6 Express the given quantity in the units in brackets.

a 124 in (ft and in)   **b** 50 ft (yd and ft)   **c** 34 yd (ft)

7 Find the size of each angle marked with a letter.

a 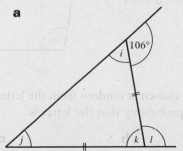   **b**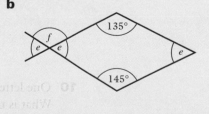

8 Which of the following shapes have rotational symmetry?

a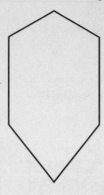

b

c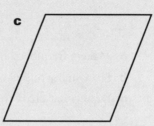

9 Construct triangle XYZ in which XY = 10.5 cm, $\widehat{X} = 65°$, $\widehat{Y} = 60°$.

10 An ordinary 6-sided dice is rolled.
What is the probability that the score is

    a 6   **b** greater than 4   **c** less than 4?

AREA

Which of these rugs takes up more floor space?

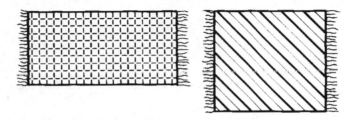

To answer this question

- We need a way to measure the amount of surface covered by each rug.
- We must use the same way to find the amount of surface for each rug so that we can compare them.

The amount of surface covered by a shape is called its *area*.
The usual way of measuring area is to find the number of squares that give the same amount of surface.

EXERCISE 14A

1 John and Hasib each traced the outline of a different leaf on squared paper. They used paper with different sized squares.

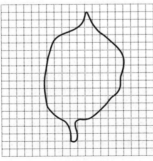

John's leaf

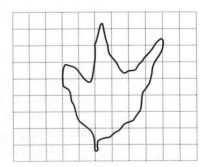

Hasib's leaf

John said that his leaf was bigger because it covered more squares than Hasib's leaf.
Discuss John's claim.

2 Beth traced the outline of a leaf on a piece of paper she found in her mother's drawer.

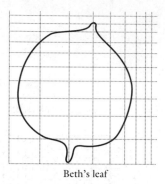

Beth's leaf

Discuss whether this can be used to estimate the area of the leaf.

3 Discuss how you would compare the areas of two different leaves.

COUNTING SQUARES TO FIND AREA

The letter L covers 7 squares. The letter C also covers 7 squares. The squares are all the same size, so we can say these two letters have the same area.

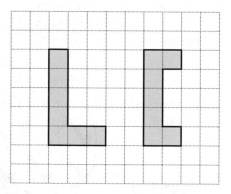

Squares are a convenient way of describing area, but there is nothing special about the size of the square we have used. If other people are going to understand what we are talking about when we say that the area of a certain shape is 12 squares, we must have a square, or unit of area, which everybody understands and which is always the same.

UNITS OF AREA

A centimetre is a standard length and a square with sides 1 cm long is said to have an area of one square centimetre. We write one square centimetre as 1 cm^2.

$\boxed{1\ \text{cm}^2}$

Other agreed lengths such as millimetres, metres and kilometres, can also be used for the sides of squares to measure area.

A square millimetre (mm^2) is a small unit of area: ▫ $1\,mm^2$

A square metre (m^2) is a fairly large unit of area. A standard single bed covers an area of about $2\,m^2$.

The unit of area used depends on what we are measuring.
We could measure the area of a small coin in square millimetres,
the area of the page of a book in square centimetres,
the area of a roof in square metres,
and the area of a county in square kilometres (km^2), or square miles if we want to use Imperial units.

We can use a square grid on a flat surface and count the squares inside the surface to estimate its area.

Sometimes the squares do not fit exactly on the area we are finding. When this is so we count a square if at least half of it is within the area we are finding, but exclude it if more than half of it is outside.

This is the outline of Hasib's leaf. It is now drawn on 1 cm squared paper.

We have ticked 20 squares, so the area of this leaf is about $20\,cm^2$.

EXERCISE 14B

This set of diagrams shows the outlines of three leaves that were drawn on 1 cm squared paper. The diagram has been reduced in size to fit this page.

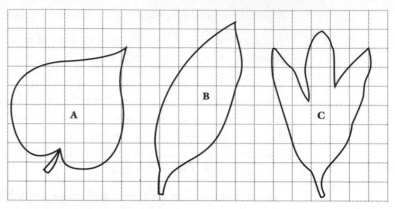

1 By counting squares find the approximate area of

   a the leaf outline marked **A**.

   b the leaf outline marked **B**.

   c the leaf outline marked **C**.

2 Which leaf has **a** the largest area **b** the smallest area?

3 Why are your answers to question **1** only approximate?

4 For each part of question **1**, decide if your answer is larger or smaller than the true area.

Which unit of area would you use to find the area of

5 this page

6 the top of a 20 p coin

7 a tennis court

8 Cornwall?

In the following questions, the figures were drawn on 1 cm squares, then the diagrams were reduced in size; so each square represents 1 cm². Find the area of the figures by counting squares.

9

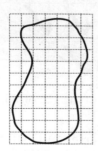

10

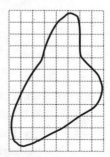

11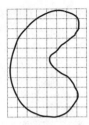

14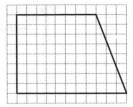

12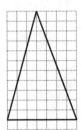

15

13

16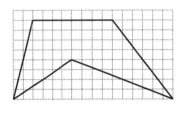

17 Which do you think is larger, 1 km² or 1 square mile? Give a reason for your answer.

18 This diagram is a map of the Isle of Wight.

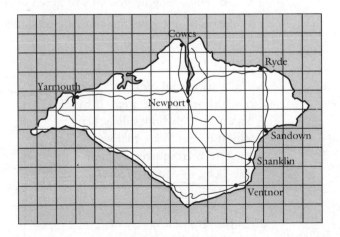

a Find, roughly, the area of the Isle of Wight as a number of grid squares.

b Each square represents 2.43 square miles. Use this to find an estimate of the area of the Isle of Wight in square miles.

19 This shows the outline of a design for a company logo drawn on 5 mm squared paper.

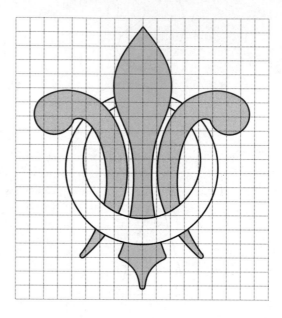

a Find the number of 5 millimetre squares covered by green.

b How many of these 5 millimetre squares make one square with an area of 1 cm^2?

c Find, roughly, the area of green in cm^2.

d Find, roughly, the area of grey in cm^2.

AREA OF A SQUARE

The square is the simplest figure of which to find the area. If we have a square whose side is 4 cm long it is easy to see that we must have 16 squares, each of side 1 cm, to cover the given square,

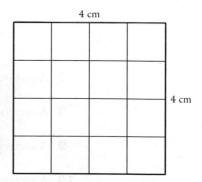

i.e. the area of a square of side 4 cm is 16 cm^2.

In general,

$$\text{area of a square } = (\text{ length of side })^2$$

AREA OF A RECTANGLE

If we have a rectangle measuring 6 cm by 4 cm we require 4 rows each containing 6 squares of side 1 cm to cover this rectangle,

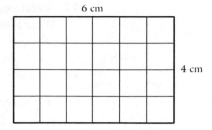

i.e. the area of the rectangle $= 6 \times 4 \, \text{cm}^2$
$$= 24 \, \text{cm}^2$$

A similar result can then be found for a rectangle of any size; for example a rectangle of length 4 cm and breadth $2\frac{1}{2}$ cm has an area of

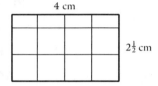

$$4 \times 2\frac{1}{2} \, \text{cm}^2$$
$$= 10 \, \text{cm}^2$$

In general, for any rectangle

$$\boxed{\text{area } = \text{ length} \times \text{breadth}}$$

EXERCISE 14C

Find the area of each of the following shapes, clearly stating the units involved.

1 A square of side 2 cm   **4** A square of side 1.5 cm

2 A square of side 8 cm   **5** A square of side 0.7 m

3 A square of side 5 cm   **6** A square of side 0.5 m

7 A rectangle measuring 5 cm by 6 cm

8 A rectangle measuring 6 cm by 8 cm

9 A rectangle measuring 3 m by 9 m

10 A rectangle measuring 1.8 mm by 2.2 mm

11 A rectangle measuring 35 km by 42 km

12 A rectangle measuring 1.5 m by 1.9 m

13 A rectangular table mat measures 24 cm by 15 cm.
What is its area?

14 Find the area of

a a soccer field measuring 110 m by 75 m

b a rugby pitch measuring 100 m by 70 m

c a tennis court measuring 26 m by 12 m.

15 A roll of wallpaper is 10 m long and 0.5 m wide.

a What is the width of the paper in metres?

b Find the area, in square metres, of the wallpaper in the roll.

c Is there enough paper in this roll to paper a wall measuring 2.4 m
by 1.9 m?
Give reasons for your answer.

**COMPOUND
FIGURES**

It is often possible to find the area of a figure by dividing it into two or
more rectangles.

EXERCISE 14D

Find the area of this tile which has been cut to shape to fit an
awkward corner.

There are two ways
to divide this shape
into two rectangles.
We could draw a line
across instead of
down as shown.

Area of A $= 6 \times 4 \, \text{cm}^2 = 24 \, \text{cm}^2$
Area of B $= 6 \times 2 \, \text{cm}^2 = 12 \, \text{cm}^2$
Therefore area of whole figure $= 24 \, \text{cm}^2 + 12 \, \text{cm}^2 = 36 \, \text{cm}^2$.

Find the areas of the following figures by dividing them into rectangles.

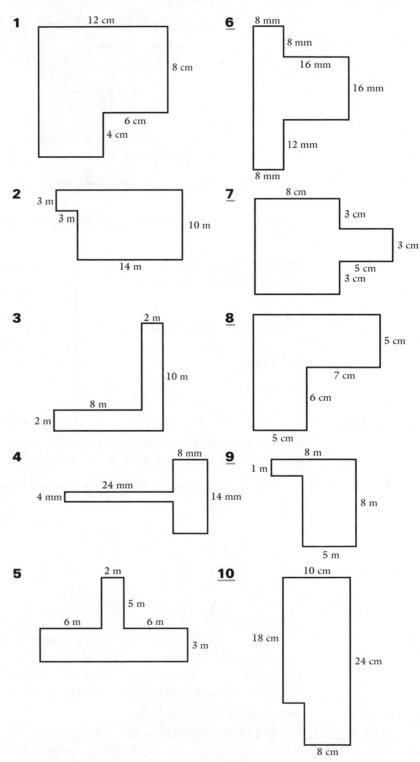

1 12 cm · 8 cm · 6 cm · 4 cm

2 3 m · 3 m · 10 m · 14 m

3 2 m · 10 m · 8 m · 2 m

4 8 mm · 24 mm · 4 mm · 14 mm

5 2 m · 5 m · 6 m · 6 m · 3 m

6 8 mm · 8 mm · 16 mm · 16 mm · 12 mm · 8 mm

7 8 cm · 3 cm · 3 cm · 5 cm · 3 cm

8 5 cm · 7 cm · 6 cm · 5 cm

9 8 m · 1 m · 8 m · 5 m

10 10 cm · 18 cm · 24 cm · 8 cm

11 The diagram shows the pieces in a puzzle. Find the area of each piece.

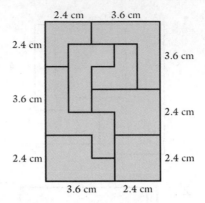

Find the area of the complete puzzle and use this to check your answers.

PERIMETER

The perimeter of a shape is the total distance round the edge of the shape.

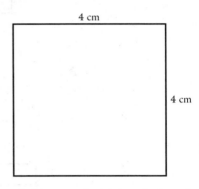

The perimeter of this square is
$4\,\text{cm} + 4\,\text{cm} + 4\,\text{cm} + 4\,\text{cm}$
i.e. $16\,\text{cm}$

If we are given a rectangle whose perimeter is 22 cm and told that the length of the rectangle is 6 cm, it is possible to find its breadth and its area.

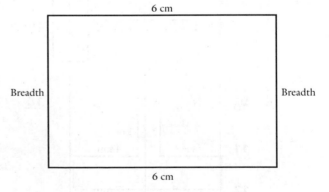

The lengths of the two longer sides add up to 12 cm so the lengths of the two shorter sides add up to $(22 - 12)\,\text{cm} = 10\,\text{cm}$

Therefore the breadth is 5 cm.

The area of this rectangle $= 6 \times 5\,\text{cm}^2$
$= 30\,\text{cm}^2$

EXERCISE 14E   Find the perimeter of each of these shapes.

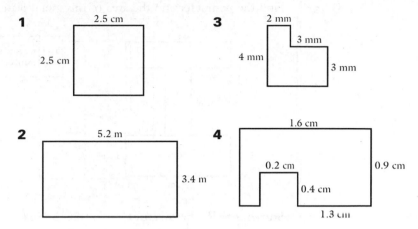

The following table gives some of the measurements of various rectangles. Fill in the values that are missing.

	Length	Breadth	Perimeter	Area
5	4 cm		12 cm	
6	5 cm		14 cm	
7		3 m	16 m	
8		6 mm	30 mm	
9	6 cm			30 cm²
10	12 m			120 m²
11		4 km		36 km²
12		7 mm		63 mm²
13		5 cm	60 cm	
14	18 cm			1680 cm²

Find the perimeter and the area of this metal plate.

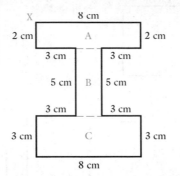

To make sure we do not miss any sides, mark one corner X.

Starting at X, the distance all round the plate and back to X is
$8 + 2 + 3 + 5 + 3 + 3 + 8 + 3 + 3 + 5 + 3 + 2$ cm $= 48$ cm.
Therefore the perimeter is **48 cm**.

The area of A $= 8 \times 2$ cm$^2 = 16$ cm^2
the area of B $= 5 \times 2$ cm$^2 = 10$ cm^2
and the area of C $= 8 \times 3$ cm$^2 = 24$ cm^2

Divide the plate into three rectangles A, B and C.

Therefore the total area $= (16 + 10 + 24)$ cm$^2 = 50$ cm^2

The following shapes are metal plates used to strengthen joints in wooden structures.
For each shape find **a** the perimeter **b** the area.

15

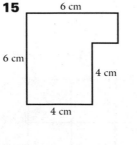

16

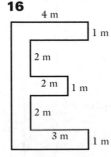

17

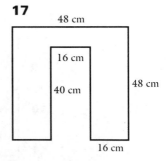

18

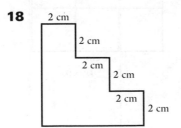

19

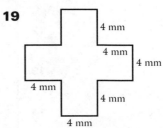

In each of the following figures find the area that is shaded.

> *Hint*: you will find it easiest to do these by subtracting the area of the missing sections from the complete area.

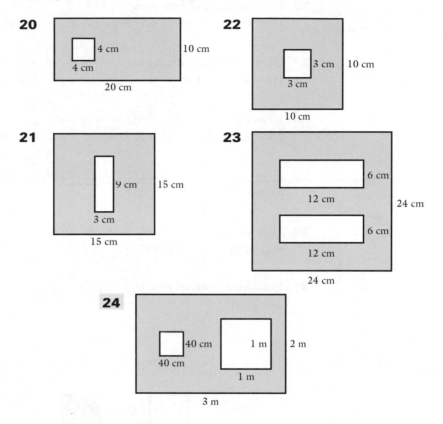

20 4 cm / 4 cm / 10 cm / 20 cm

22 3 cm / 3 cm / 10 cm / 10 cm

21 9 cm / 3 cm / 15 cm / 15 cm

23 6 cm / 12 cm / 6 cm / 12 cm / 24 cm / 24 cm

24 40 cm / 40 cm / 1 m / 1 m / 2 m / 3 m

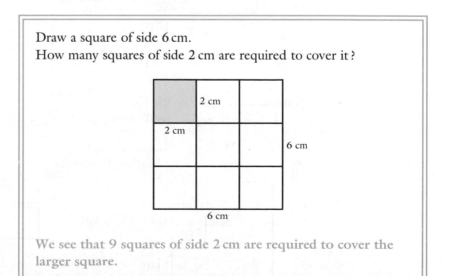

Draw a square of side 6 cm.
How many squares of side 2 cm are required to cover it?

2 cm / 2 cm / 6 cm / 6 cm

We see that 9 squares of side 2 cm are required to cover the larger square.

25 Draw a square of side 4 cm.
How many squares of side 2 cm are required to cover it?

26 Draw a square of side 9 cm.
How many squares of side 3 cm are required to cover it?

27 Draw a rectangle measuring 6 cm by 4 cm.
How many squares of side 2 cm are required to cover it?

28 Draw a rectangle measuring 9 cm by 6 cm.
How many squares of side 3 cm are required to cover it?

29 How many squares of side 5 cm are required to cover a rectangle measuring 45 cm by 25 cm?

30 How many squares of side 4 cm are required to cover a rectangle measuring 1 m by 80 cm?

31 How many squares of side 1 mm are required to cover a square of side 1 cm?

CHANGING UNITS OF AREA

A square of side 1 cm may be divided into 100 squares of side 1 mm,

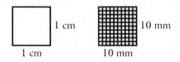

i.e. $$1\,\text{cm}^2 = 100\,\text{mm}^2$$

Similarly, since 1 m = 100 cm,
    1 square metre = 100×100 square centimetres,

i.e. $$1\,\text{m}^2 = 10\,000\,\text{cm}^2$$

and, as 1 km = 1000 m,
1 square kilometre = 1000×1000 square metres,

i.e. $$1\,\text{km}^2 = 1\,000\,000\,\text{m}^2$$

When we convert from a unit of area that is large to a unit of area that is smaller, we must remember that the *number* of units will be bigger,

e.g. $2 \, km^2 = 2 \times 1\,000\,000 \, m^2$
$= 2\,000\,000 \, m^2$

but if we convert from a unit of area that is small into a larger unit of area, the number of units will be smaller,

e.g. $500 \, mm^2 = 500 \div 100 \, cm^2$
$= 5 \, cm^2$

EXERCISE 14F

> Express $5 \, m^2$ in cm^2.
>
> As $1 \, m^2 = 10\,000 \, cm^2$,
> $5 \, m^2 = 5 \times 10\,000 \, cm^2$
> $= 50\,000 \, cm^2$

1 Express in cm^2
   a $3 \, m^2$ **b** $12 \, m^2$ **c** $7.5 \, m^2$ **d** $82 \, m^2$ **e** $0.5 \, m^2$

2 Express in mm^2
   a $14 \, cm^2$ **b** $3 \, cm^2$ **c** $7.5 \, cm^2$ **d** $26 \, cm^2$ **e** $3.2 \, cm^2$

3 Express $0.056 \, m^2$ in cm^2.

4 Express in cm^2
   a $400 \, mm^2$ **c** $50 \, mm^2$ **e** $734 \, mm^2$
   b $2500 \, mm^2$ **d** $25 \, mm^2$ **f** $1220 \, mm^2$

5 Express in m^2
   a $5500 \, cm^2$ **c** $760 \, cm^2$ **e** $29\,700 \, cm^2$
   b $140\,000 \, cm^2$ **d** $18\,600 \, cm^2$ **f** $192\,000 \, cm^2$

6 Express in km^2
   a $7\,500\,000 \, m^2$ **c** $50\,000 \, m^2$ **e** $176 \, m^2$
   b $430\,000 \, m^2$ **d** $245\,000 \, m^2$ **f** $750\,000 \, m^2$

7 Use 5 miles $\approx$ 8 km to find an approximate conversion from square miles to square kilometres.

In many problems involving finding the area of a rectangle, the length and breadth are given in different units. When this is so we must change the units so that all the measurements are in the same units.

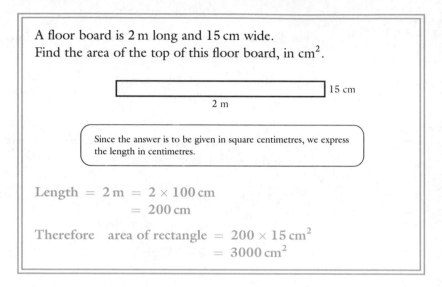

A floor board is 2 m long and 15 cm wide.
Find the area of the top of this floor board, in cm^2.

15 cm
2 m

Since the answer is to be given in square centimetres, we express the length in centimetres.

$$\text{Length} = 2\,\text{m} = 2 \times 100\,\text{cm}$$
$$= 200\,\text{cm}$$

$$\text{Therefore} \quad \text{area of rectangle} = 200 \times 15\,\text{cm}^2$$
$$= 3000\,\text{cm}^2$$

Find the area of each of the following rectangles, giving your answer in the unit in brackets.

	Length	Breadth	
8	10 m	50 cm	(cm^2)
9	6 cm	30 mm	(mm^2)
10	50 m	35 cm	(cm^2)
11	140 cm	1 m	(cm^2)
12	400 cm	200 cm	(m^2)
13	3 m	$\frac{1}{2}$ m	(cm^2)
14	1.2 cm	5 mm	(mm^2)
15	0.4 km	0.05 km	(m^2)

16 The top of my desk is 1.5 m long and 60 cm wide.
Find its area in cm^2.

17 A roll of wallpaper is 10 m long and 50 cm wide.
Find its area in square metres.

18 A school hall measuring 20 m by 15 m is to be covered with square floor tiles of side 50 cm.
How many tiles are required?

19 A rectangular carpet measures 4 m by 3 m.
Find its area.
What is the cost of cleaning this carpet at 75 p per square metre?

20 How many square linen serviettes, of side 50 cm, may be cut from a roll of linen 25 m long and 1 m wide?

21 How many square concrete paving slabs, each of side 0.5 m, are required to pave a rectangular yard measuring 9 m by 6 m?

22 A patio that is 9 m square is to be covered with 450 mm square paving stones and edged, all the way round, with 250 mm long bricks.
How many paving stones are needed?
What extra information do you need in order to find the number of bricks needed?

23 A rectangular field measuring 40 m by 25 m is to be sown with grass seed so that 90 grams of seed is used to cover each square metre.
Each box of seeds holds 5 kg.
How many boxes are needed?

24 A rectangular lawn measuring 3 m by 4 m has a 1 m wide path round the outside edge.
Find the area of this path.

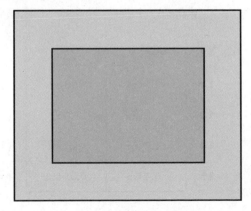

25 Diana is tiling a wall with two different sized square tiles. The smaller tile has sides 64 mm long. Four of these smaller tiles cover the same area as one larger tile.
What is the length of a side of a larger tile?

INVESTIGATION

Shapes made with 1 cm squares

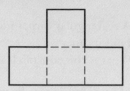

a This shape has a perimeter of 10 cm.
What is its area?

b Find other shapes made from 1 cm squares that also have a perimeter of 10 cm.
Which one has the largest area?

c Find other shapes that have the same area as the shape above.
Which shape has the shortest perimeter?

d Investigate different shapes with a perimeter of 16 cm.
Find the shape with the largest possible area.

e Investigate different shapes with an area of 6 cm².
Which shape has the shortest perimeter?

f For a given area, what shape has the shortest perimeter?

g A rectangle has the same number of square centimetres of area as it has centimetres of perimeter.
Find possible whole number values for the length and breadth of this rectangle. (There are two different rectangles with this property.)

PARALLEL LINES

15

Two straight lines that are always the same distance apart, however far they are drawn, are called parallel lines.

The lines in your exercise books are parallel. You can probably find many other examples of parallel lines; for example, the edges on many buildings.

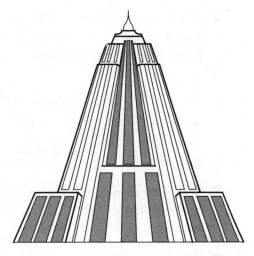

1 Discuss with your group examples of parallel lines that

a are horizontal and you can see them in the room

b are vertical and you can see them in the room

c do not look parallel

2 Using the lines in your exercise book, draw three lines that are parallel. Do not make them all the same distance apart. For example

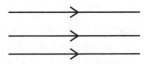

(We use arrows to mark lines that are parallel.)

3 Using the lines in your exercise book, draw two parallel lines. Make them fairly far apart. Now draw a slanting line across them. For example

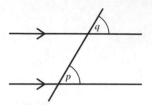

Mark the angles in your drawing that are in the same position as those in the diagram.
Are they acute or obtuse angles?
Measure your angles marked p and q.

4 Draw a grid of parallel lines like the diagram below. Use the lines in your book for one set of parallels and use the two sides of your ruler to draw the slanting parallels.

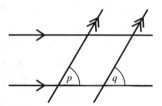

Mark your drawing like the diagram.
Are your angles p and q acute or obtuse?
Measure your angles p and q.

5 Repeat question **4** but change the direction of your slanting lines.

6 Draw three slanting parallel lines like the diagram below with a horizontal line cutting them. Use the two sides of your ruler and move it along to draw the third parallel line.

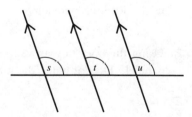

Mark your drawing like the diagram.
Decide whether angles s, t and u are acute or obtuse and then measure them.

7 Repeat question **6** but change the slope of your slanting lines.

CORRESPONDING
ANGLES

In the last exercise, lines were drawn that crossed a set of parallel lines.

A line that crosses a set of parallel lines is called a *transversal*.

When you have drawn several parallel lines you should notice that

two parallel lines on the same flat surface will never meet however far they are drawn

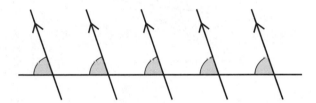

If you draw the diagram above by moving your ruler along you can see that all the shaded angles are equal. These angles are all in corresponding positions: they are all above the transversal and to the left of the parallel lines. Angles like these are called *corresponding angles*.

When two parallel lines are cut by a transversal, the corresponding angles are equal.

EXERCISE 15B

In the diagrams below write down the angle that corresponds to the shaded angle.

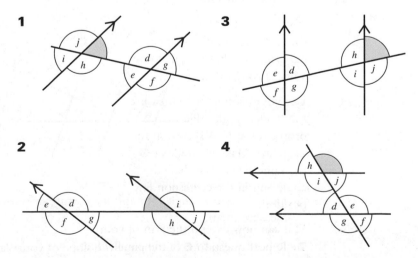

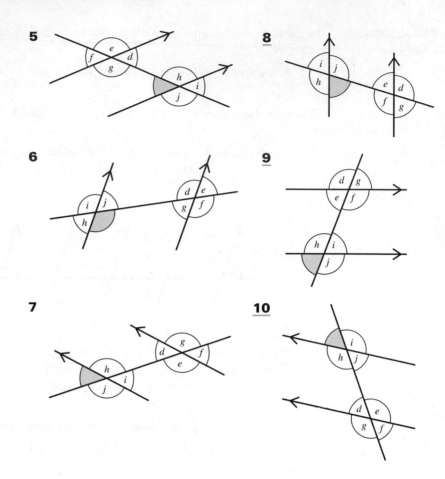

5

6

7

8

9

10

DRAWING PARALLEL LINES (USING A PROTRACTOR)

The fact that the corresponding angles are equal gives us a method for drawing parallel lines.

If you need to draw a line through the point C that is parallel to the line AB, first draw a line through C to cut AB.

Use your protractor to measure the shaded angle. Place your protractor at C as shown in the diagram. Make an angle at C the same size as the shaded angle and in the corresponding position.

You can now extend the arm of your angle both ways, to give the parallel line.

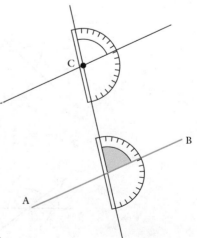

EXERCISE 15C

1 Using your protractor draw a grid of parallel lines like the one in the diagram. (It does not have to be an exact copy.)

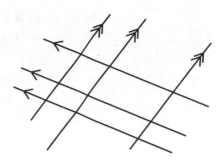

2 Trace the diagram below.

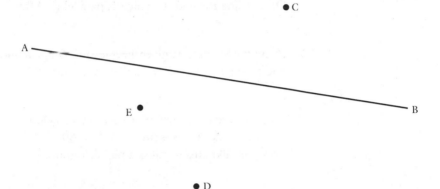

Now draw lines through the points C, D and E so that each line is parallel to AB.

3 Draw a sloping line in your exercise book. Mark a point C above the line. Use your protractor to draw a line through C parallel to your first line.

4 Trace the diagram below.

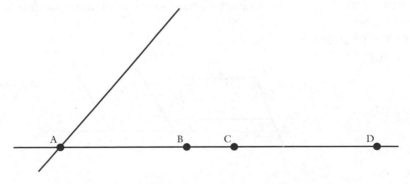

Measure the acute angle at A. Draw the corresponding angles at B, C and D. Extend the arms of your angles so that you have a set of four parallel lines.

In questions **5** to **8** remember to draw a rough sketch before doing the accurate drawing.

5 Draw an equilateral triangle with sides each 8 cm long. Label the corners A, B and C.
Draw a line through C that is parallel to the side AB.

6 Draw an isosceles triangle ABC with base AB which is 10 cm long and base angles at A and B which are each 30°.
Draw a line through C which is parallel to AB.

7 Draw the triangle as given in question **6** again and this time draw a line through A which is parallel to the side BC.

8 Make an accurate drawing of the figure below where the side AB is 7 cm, the side AD is 4 cm and $\widehat{A} = 60°$.
(A figure like this is called a parallelogram.)

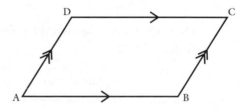

PROBLEMS INVOLVING CORRESPONDING ANGLES

The simplest diagram for a pair of corresponding angles is an F shape.

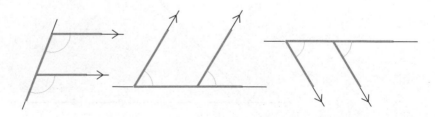

Looking for an F shape may help you to recognise the corresponding angles.

EXERCISE 15D

Write down the size of the angle marked *d*.

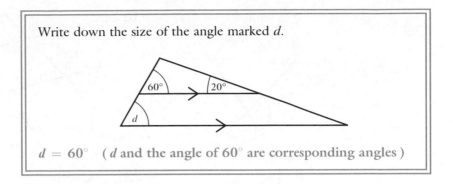

$d = 60°$ (*d* and the angle of 60° are corresponding angles)

Write down the size of the angle marked *d* in each of the following diagrams.

1

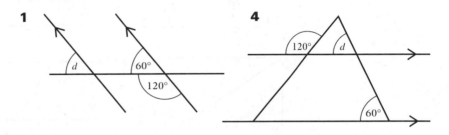

4

2

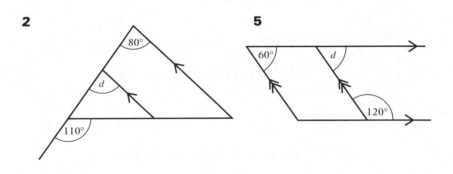

5

3

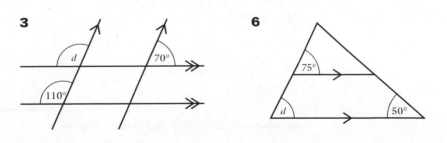

6

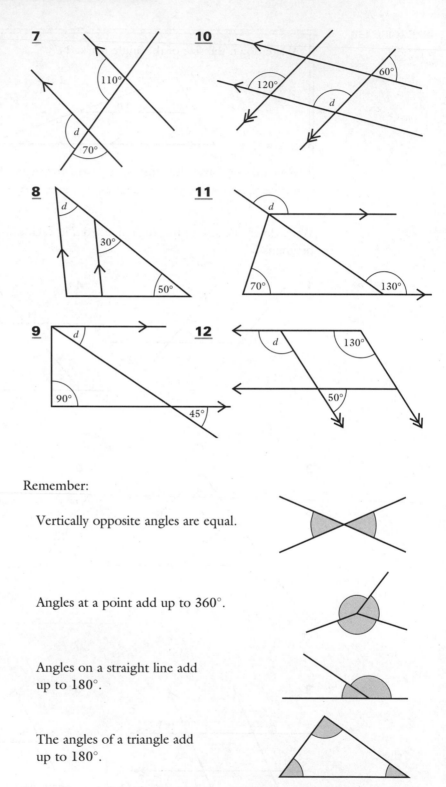

Remember:

Vertically opposite angles are equal.

Angles at a point add up to 360°.

Angles on a straight line add up to 180°.

The angles of a triangle add up to 180°.

You will need these facts in the next exercise.

EXERCISE 15E

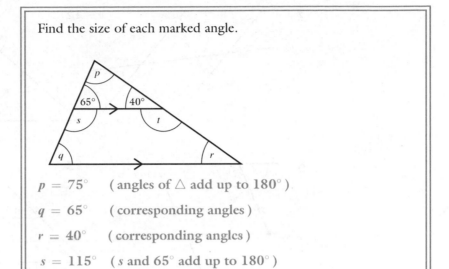

Find the size of each marked angle.

$p = 75°$ (angles of △ add up to 180°)

$q = 65°$ (corresponding angles)

$r = 40°$ (corresponding angles)

$s = 115°$ (s and 65° add up to 180°)

$t = 140°$ (t and 40° add up to 180°)

Find the size of each marked angle. Give reasons for any statements that you make.

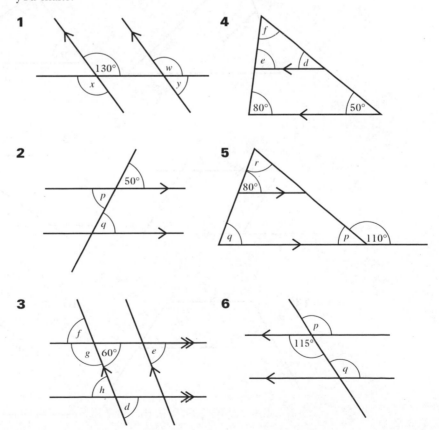

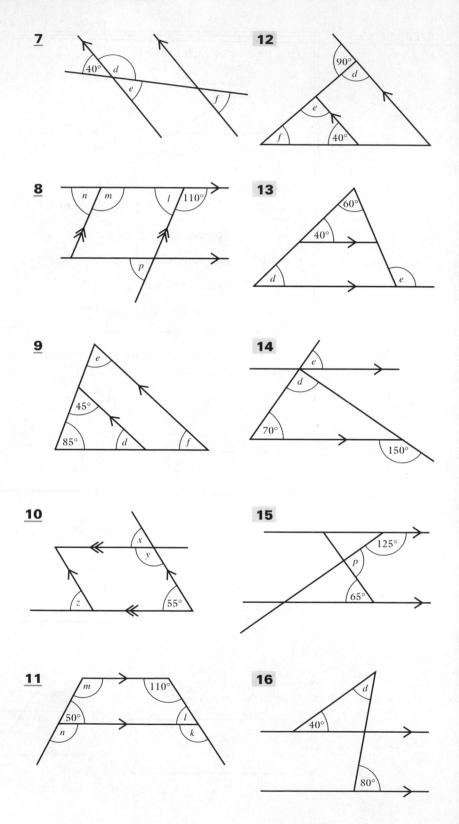

7

8

9

10

11

12

13

14

15

16

Find the size of angle *d* in questions **17** to **24**.

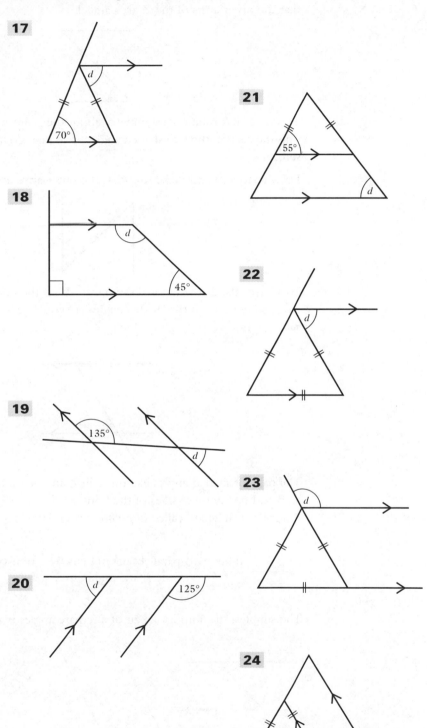

17

18

19

20

21

22

23

24

ALTERNATE
ANGLES

Draw a large letter Z. Use the lines of your exercise book to make sure that the outer arms of the Z are parallel.

This letter has rotational symmetry about the point marked with a cross. This means that the two shaded angles are equal. Measure them to make sure.

Draw a large N and make sure that the outer arms are parallel.

This letter also has rotational symmetry about the point marked with a cross, so once again the shaded angles are equal. Measure them to make sure.

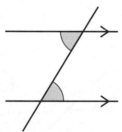

The pairs of shaded angles like those in Z and N are between the parallel lines and on alternate sides of the transversal.
Angles like these are called *alternate angles*.

> When two parallel lines are cut by a transversal, the alternate angles are equal.

The simplest diagram for a pair of alternate angles is a Z shape.

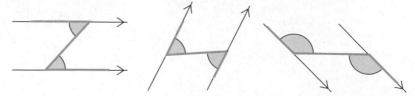

Looking for a Z shape may help you to recognise the alternate angles.

EXERCISE 15F   Write down the angle which is alternate to the shaded angle in the
following diagrams.

1

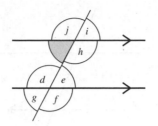

6

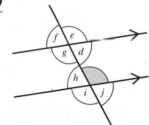

2

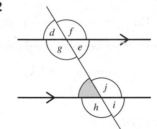

7

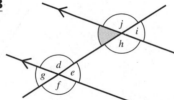

3

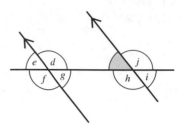

8

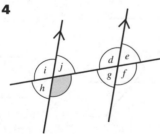

4

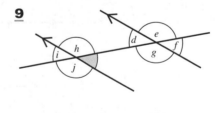

9

5

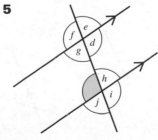

10

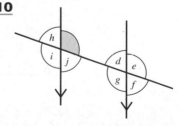

PROBLEMS INVOLVING ALTERNATE ANGLES

Without doing any measuring we can show that alternate angles are equal by using the facts that we already know:

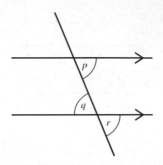

$p = r$ because they are corresponding angles

$q = r$ because they are vertically opposite angles

$\therefore \; p = q$ and these are alternate angles

EXERCISE 15G

Find the size of each marked angle.

1

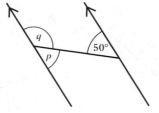

4

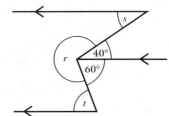

2

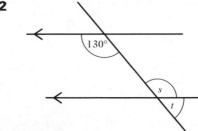

5

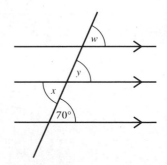

3

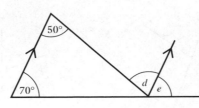

6

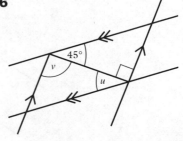

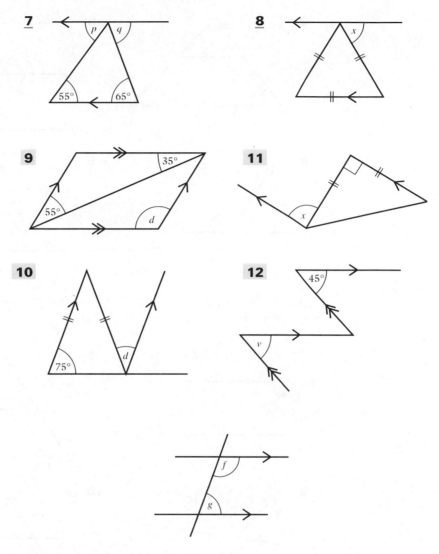

INTERIOR
ANGLES

In the diagram above, f and g are on the same side of the transversal and 'inside' the parallel lines.

Pairs of angles like f and g are called *interior angles*.

EXERCISE 15H   In each of the following diagrams, two of the marked angles are a pair of interior angles. Name them.

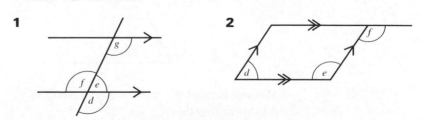

3

5

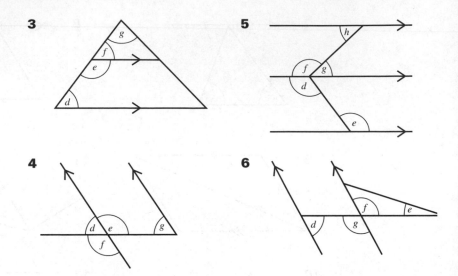

4

6

In the following diagrams, use the information given to find the size of p and of q. Then find the sum of p and q.

7

9

8

10

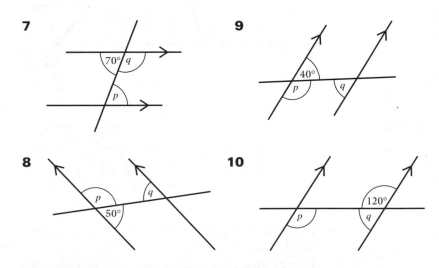

11 Make a large copy of the diagram below. Use the lines of your book to make sure that the outer arms of the 'U' are parallel.

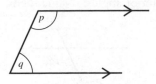

Measure each of the interior angles p and q.
Add them together.

The sum of a pair of interior angles is 180°

You will probably have realised this fact by now. We can show that it is true from the following diagram.

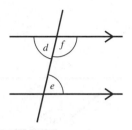

$d + f = 180°$ because they are angles on a straight line

$d = e$ because they are alternate angles

So $e + f = 180°$

The simplest diagram for a pair of interior angles is a U shape.

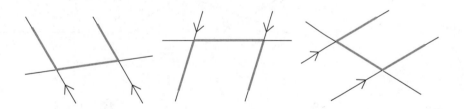

Looking for a U shape may help you to recognise a pair of interior angles.

EXERCISE 15I   Find the size of each marked angle.

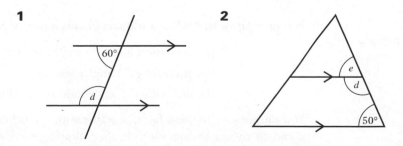

1

2

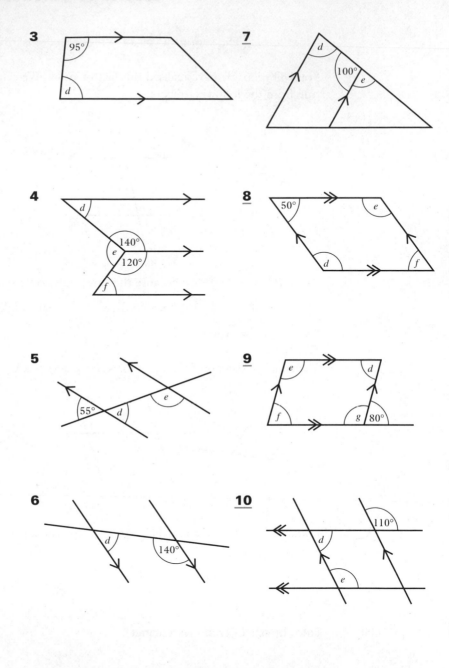

MIXED EXERCISES

You now know that when a transversal cuts a pair of parallel lines

the corresponding (F) angles are equal

the alternate (Z) angles are equal

the interior (U) angles add up to 180°.

You can use any of these facts, together with the other angle facts you know, to answer the questions in the following exercises.

EXERCISE 15J Find the size of each marked angle.

1

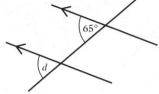

5

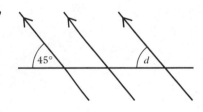

2

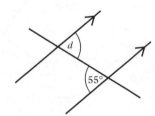

6

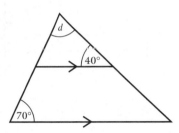

3

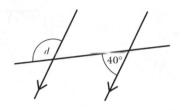

7

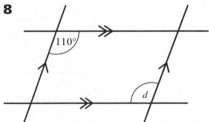

4

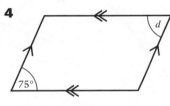

8

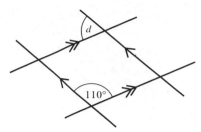

9 Construct a triangle ABC in which AB = 12 cm, BC = 8 cm and
AC = 10 cm.
Find the midpoint of AB and mark it D.
Find the midpoint of AC and mark it E.
Join ED.
Measure AD̂E and AB̂C.
(AD̂E means the angle at D formed by the lines AD and DE.)
What can you say about the lines DE and BC?

EXERCISE 15K   Find the size of each marked angle.

1

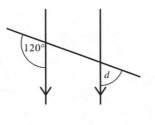

2

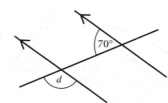

3

4

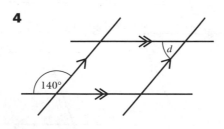

5

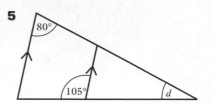

6

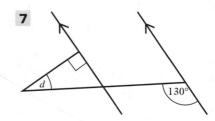

7

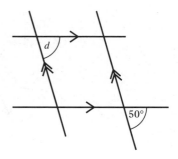

8

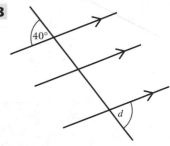

9 Construct this parallelogram, making it full size.

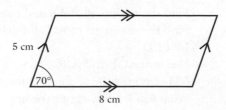

5 cm

70°

8 cm

INVESTIGATION

This diagram represents a child's billiards table.

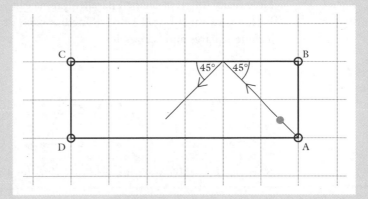

There is a pocket at each corner.
The ball is projected from the corner A at 45° to the sides of the table.
It carries on bouncing off the sides at 45° until it goes down a pocket.
(This is a very superior toy – the ball does not lose speed however
many times it bounces !)

a How many bounces are there before the ball goes down a pocket ?

b Which pocket does it go down ?

c What happens if the table is 2 squares by 8 squares ?

d Can you predict what happens for a 2 by 20 table ?

e Now try a 2 by 3 table.

f Investigate for other sizes of tables. Start by keeping the width at
2 squares. Then try other widths. Copy this table and fill in the
results.

Size of table	Number of bounces	Pocket
2 × 6		
2 × 8		
2 × 3		
2 × 5		

g Can you predict what happens with a 3 × 12 table ?

Copy this grid.

How many different-shaped parallelograms can you draw on this grid? Each vertex must be on a dot. One has been drawn for you. Do not include squares and rectangles.

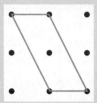

Now try the problem with a grid of 4 dots by 4 dots.

COORDINATES

Your parents tell you that they are going to Brixworth to see some distant relatives. You have no idea where Brixworth is but you can find out by using a road atlas.
When you turn up the name in the Index you find

56   Brixworth F5

This is a short description of where to find Brixworth on the map.

Do you know what it means?

EXERCISE 16A

1 Discuss different ways in which you can give the position of a place on a map so that another person can find it.

2 Look up the index in several different atlases. How many different ways are there of helping you to find the position of a place on a map?

3 A friend rings you up asking what way the pieces can move in a game of chess.
Discuss how you can describe the ways in which the different chess pieces can move.

4 Can you think of other situations where you need a short way of describing the position of a place?

PLOTTING POINTS USING POSITIVE COORDINATES

There are many occasions when you need to describe the position of an object. For example, you may want to tell a friend how to find your house, or you might want to find a square in the game of battleships. An air traffic controller may have to describe the position of an aeroplane showing up on a radar screen. In mathematics we need a quick way to describe the position of a point.

We do this by using squared paper and marking a point O at the corner of one square. We then draw a line through O across the page. This line is called O*x*. Next we draw a line through O up the page. This line is called O*y*. Starting from O we then mark numbered scales on each line.

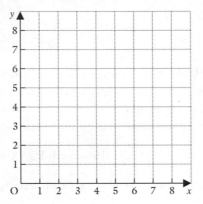

O is called the origin

O*x* is called the *x*-axis

O*y* is called the *y*-axis.

We can now describe the position of a point A as follows:

start from O and move 3 squares along O*x*,
then move 5 squares up from O*x*.

We always use the same method to describe the position of a point:

start from O, *first* move *along* and *then up*.

We can now shorten the description of the position of the point A to the number pair (3, 5).

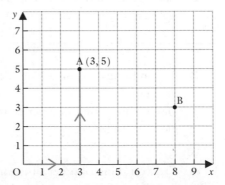

The number pair (3, 5) gives the *coordinates* of A.
The first number, 3, is called the *x*-coordinate of A.
The second number, 5, is called the *y*-coordinate of A.

Now consider another point B whose *x*-coordinate is 8
and whose *y*-coordinate is 3.

If we simply refer to the point B (8, 3) this tells us all that we need to know about the position of B.

The origin is the point (0, 0).

EXERCISE 16B

1 Write down the coordinates of the points A, B, C, D, E, F, G and H.

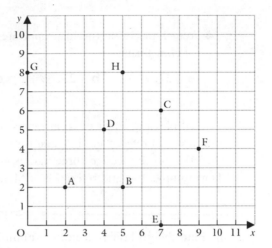

Use this map for questions 2 and 3.

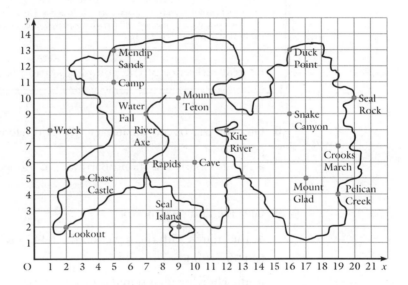

2 Which places on the map have the following pairs of coordinates?

a (2, 2)   **c** (16, 13)   **e** (9, 10)

b (19, 4)   **d** (3, 5)   **f** (7, 6)

3 Give the pair of coordinates for each of the following places.

a Seal Island   **e** Mount Glad

b Pelican Creek   **f** The source of the Kite River

c Snake Canyon   **g** The point where the Kite River enters the sea

d Mendip Sands   **h** Water Fall

Questions **4**, **5** and **6** use the map on the opposite page.

4 Which places on the map opposite are given by the following pairs of coordinates?

   a $(27, 36)$ **b** $(29, 39)$ **c** $(25, 39)$ **d** $(28, 34)$

Instead of using a pair of coordinates, a place can be described by a 6-figure grid reference. The first three figures give the position across the grid and the second three figures give the position up the grid. For example, the grid reference of Manor Farm is 255 352.

Find the grid reference of the pumping station, near the tunnel.

> We can use the divisions of the squares to give the position of the pumping station. It is between the vertical grid lines marked 25 and 26 and is $\frac{4}{10}$ of the side of the square from the line marked 25, so the horizontal reference is 254. It is on the grid line marked 36, so its vertical grid reference is 360.

The grid reference of the pumping station is 254 360

---

5 Give the pair of coordinates and the grid references that fix the position of
   a Sands Court   **b** Long Barrow   **c** Abbey Gate

6 Give the grid reference of

   a the church with the tower on the B442 (▮)

   b Horse Bridge

   c the church with a spire (●)

   d Horton Hall

7 Draw a set of axes of your own.
Give them scales from 0 to 10.
Mark the following points and label each point with its own letter.

$A(2, 8)$ $B(4, 9)$ $C(7, 9)$ $D(8, 7)$ $E(8, 6)$ $F(9, 4)$
$G(8, 4)$ $H(7, 3)$ $I(5, 3)$ $J(7, 2)$ $K(7, 1)$ $L(4, 2)$
$M(2, 0)$ $N(0, 2)$

Now join your points together in alphabetical order and join A to N.

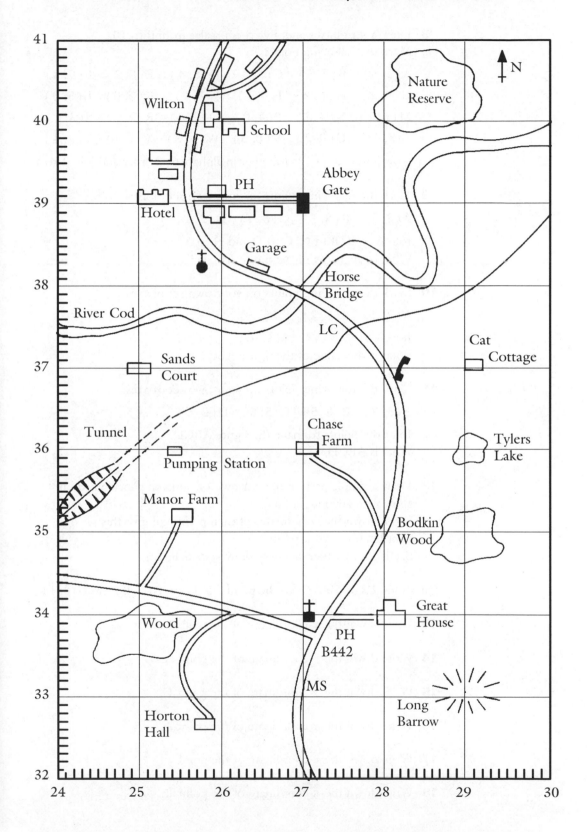

8 Draw a set of axes and give them scales from 0 to 10.
Mark the following points.

$A(2,5)$ $B(7,5)$ $C(7,4)$ $D(8,4)$ $E(8,3)$ $F(9,3)$
$G(9,2)$ $H(6,2)$ $I(6,1)$ $J(7,1)$ $K(7,0)$ $L(5,0)$
$M(5,2)$ $N(4,2)$ $P(4,0)$ $Q(2,0)$ $R(2,1)$ $S(3,1)$
$T(3,2)$ $U(0,2)$ $V(0,3)$ $W(1,3)$ $X(1,4)$ $Y(2,4)$

Now join your points together in alphabetical order and join A to Y.

9 Mark the following points on your own set of axes.

$A(2,7)$ $B(8,7)$ $C(8,1)$ $D(2,1)$

Join A to B, B to C, C to D and D to A.
What is the name of the figure ABCD?

10 Mark the following points on your own set of axes.

$A(2,2)$ $B(8,2)$ $C(5,5)$

Join A to B, B to C and C to A.
What is the name of the figure ABC?

11 Mark the following points on your own set of axes.

$A(5,2)$ $B(8,5)$ $C(5,8)$ $D(2,5)$

Joint the points to make the figure ABCD.
What is ABCD?

12 Draw a simple pattern of your own on squared paper but do not
show it to anyone.
Write down the coordinates of each point and give this set of
coordinates to your partner.
See if your partner can now draw your diagram.

Questions **13** to **18** refer to the points $A(1,7)$, $B(5,0)$ and $C(0,14)$.

13 Write down the x-coordinate of the point B.

14 Write down the y-coordinate of the point A.

15 Write down the x-coordinate of the point C.

16 Write down the x-coordinate of the point A.

17 Write down the y-coordinate of the point C.

18 Write down the y-coordinate of the point B.

Questions **19** to **24** refer to points in the following diagram.

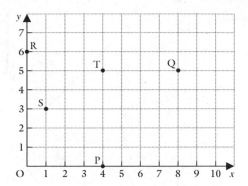

19 Write down the *y*-coordinate of the point T.

20 Write down the *x*-coordinate of the point P.

21 Write down the *x*-coordinate of the point S.

<u>**22**</u> Write down the *y*-coordinate of the point R.

<u>**23**</u> Write down the *y*-coordinate of the point Q.

<u>**24**</u> Write down the *x*-coordinate of the point R.

For each of the following questions you will need to draw your own set of axes.

25 The points A(2, 1), B(6, 1) and C(6, 5) are three corners of a square ABCD.
Mark the points A, B and C.
Find the point D and write down the coordinates of D.

26 The points A(2, 1), B(2, 3) and C(7, 3) are three vertices of a rectangle ABCD.
Mark the points and find the point D.
Write down the coordinates of D.

27 The points A(1, 4), B(4, 7) and C(7, 4) are three vertices of a square ABCD.
Mark the points A, B and C and find D.
Write down the coordinates of D.

Questions **28** to **31** refer to the following diagram.

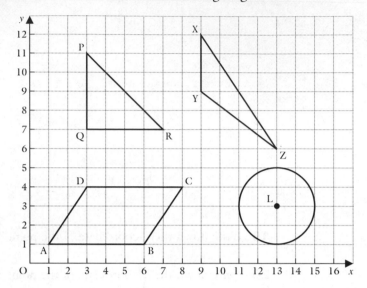

28 Write down the coordinates of the vertices X, Y and Z of triangle XYZ.

29 Write down the coordinates of the vertices of the isosceles triangle PQR.
Write down the lengths of the two equal sides as a number of sides of squares.

30 Write down the coordinates of the vertices of the parallelogram ABCD.
How long is AB?
How long is DC?

31 Write down the coordinates of the centre, L, of the circle.
What is the diameter of this circle?

You will need to draw your own set of axes for these questions.

32 Mark the points A (2, 4) and B (8, 4).
Join A to B and find the point C which is the midpoint (the exact middle) of the line AB.
Write down the coordinates of C.

33 Mark the points P (3, 5) and Q (3, 9).
Join P and Q and mark the point R which is the midpoint of PQ.
Write down the coordinates of R.

34 Mark the points A (0, 5) and B (4, 1).
Find the coordinates of the midpoint of AB.

QUADRILATERALS

A quadrilateral has four sides.
No two of the sides need be equal
and no two of the sides need be
parallel.

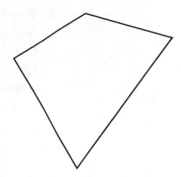

There are, however, some special quadrilaterals, such as a square, which
have some sides parallel and/or some sides equal.

The questions in the next exercise investigate the properties of these
special quadrilaterals.

EXERCISE 16C

If you are not sure whether two lines are equal, *measure them.*

If you are not sure whether two lines are parallel, *measure a pair of corresponding angles.*

1 The Square
A(3, 2), B(11, 2), C(11, 10) and D(3, 10) are the four corners
of a square.
Using squared paper mark these points on your own set of axes and
then draw the square ABCD.

a Write down, as a number of sides of squares, the lengths of the
sides AB, BC, CD and DA.

b Which side is parallel to AB ?
Are BC and AD parallel ?

c What is the size of each angle of the square ?

2 The Rectangle
A(2, 2), B(2, 7), C(14, 7) and D(14, 2) are the vertices of a
rectangle ABCD.
Draw the rectangle ABCD on your own set of axes.

a Write down the sides which are equal in length.

b Write down the pairs of sides which are parallel.

c What is the size of each angle of the rectangle ?

3 The Rhombus

A (8, 1), B (11, 7), C (8, 13) and D (5, 7) are the vertices of a rhombus ABCD.

Draw the rhombus on your own set of axes.

a Write down the sides which are equal in length.

b Write down the pairs of sides which are parallel.

c Measure the angles of the rhombus.
Are any of the angles equal?

4 The Parallelogram

A (2, 2), B (14, 2), C (17, 7) and D (5, 7) are the vertices of a parallelogram.

Draw the parallelogram on your own set of axes.

a Write down which sides are equal in length.

b Write down which sides are parallel.

c Measure the angles of the parallelogram.
Write down which, if any, of the angles are equal.

5 The Trapezium

A (1, 1), B (12, 1), C (10, 5) and D (5, 5) are the vertices of a trapezium.

Draw the trapezium on your own set of axes.

a Write down which, if any, of the sides are the same length.

b Write down which, if any, of the sides are parallel.

c Write down which, if any, of the angles are equal.

PROPERTIES OF THE SIDES AND ANGLES OF THE SPECIAL QUADRILATERALS

We can summarise our investigations in the last exercise as follows:

In a square
- all four sides are the same length
- both pairs of opposite sides are parallel
- all four angles are right angles.

In a rectangle
- both pairs of opposite sides are the same length
- both pairs of opposite sides are parallel
- all four angles are right angles.

In a rhombus
- all four sides are the same length
- both pairs of opposite sides are parallel
- the opposite angles are equal.

In a parallelogram
- the opposite sides are the same length
- the opposite sides are parallel
- the opposite angles are equal.

In a trapezium
- just one pair of opposite sides are parallel.

EXERCISE 16D

In the following questions the points A, B, C and D are the vertices of a quadrilateral.

Draw the figure ABCD on your own set of axes and write down which type of quadrilateral it is.

1 A(2,4) B(7,4) C(8,7) D(3,7)

2 A(2,2) B(6,0) C(7,2) D(3,4)

3 A(2,2) B(7,2) C(5,5) D(3,5)

4 A(2,0) B(6,0) C(6,4) D(2,4)

5 A(1,1) B(4,0) C(4,6) D(1,3)

6 A(3,1) B(6,3) C(3,5) D(0,3)

7 A(1,3) B(4,1) C(6,4) D(3,6)

8 A(2,4) B(3,7) C(9,5) D(8,2)

9 A(3,1) B(5,1) C(3,5) D(1,5)

10 A(0,0) B(5,0) C(8,4) D(3,4)

NEGATIVE COORDINATES

If A(2,0), B(4,2) and C(6,0) are three corners of a square ABCD, we can see that the fourth corner, D, is two squares below the x-axis.

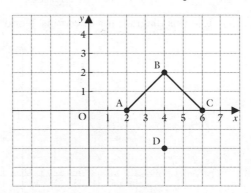

To describe the position of D we need to extend the scale on the y-axis below zero. To do this we use the numbers $-1, -2, -3, -4, \ldots$. These are called *negative* numbers.

In the same way we can use the negative numbers $-1, -2, -3, \ldots$ to extend the scale on the x-axis to the left of zero.

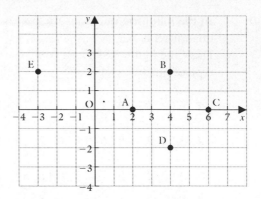

The y-coordinate of the point D is written -2 and is called 'negative 2'.

Some people write the negative sign higher up, e.g. $^-2$, to distinguish 'negative 2' from 'take away 2'.

The x-coordinate of the point E is written -3 and is called 'negative 3'.

The numbers $1, 2, 3, 4 \ldots$ are called positive numbers. They could be written as $+1, +2, +3, +4, \ldots$ but we do not usually put the $+$ sign in.

Now   D is 4 squares to the right of O so its x-coordinate is 4

    and   2 squares below the x-axis so its y-coordinate is -2,

                      D is the point $(4, -2)$

E is 3 squares to the left of O so its x-coordinate is -3

and 2 squares up from O so its y-coordinate is 2,

                      E is the point $(-3, 2)$

EXERCISE 16E   Use this diagram for questions **1** and **2**.

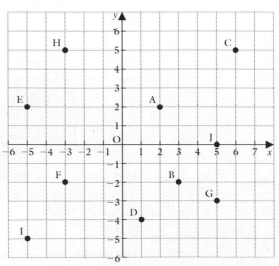

1 Write down the *x*-coordinate of each of the points A, B, C, D, E, F, G, H, I, J and O (the origin).

2 Write down the *y*-coordinate of each of the points A, B, C, D, E, H, I and J.

The point Q has a *y*-coordinate of -10.
How many squares above or below the *x*-axis is the point Q?

Q is 10 squares below the *x*-axis.

How many squares above or below the *x*-axis is each of the following points?

3 P: the *y*-coordinate is -5

5 B: the *y*-coordinate is 10

4 L: the *y*-coordinate is $+3$

6 A: the *y*-coordinate is 0

How many squares to the left or to the right of the *y*-axis is each of the following points?

7 Q: the *x*-coordinate is 3

9 S: the *x*-coordinate is -7

8 R: the *x*-coordinate is -5

10 V: the *x*-coordinate is 0

Write down the coordinates
of the point A.

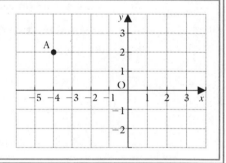

A is the point ($-4, 2$).

---

11 Write down the
coordinates of the
points A, B, C, D, E,
F, G, H, I and J.

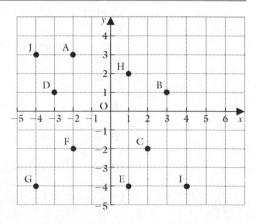

In questions **12** to **16** draw your own set of axes and mark a scale on each one from −5 to 5.

12 Mark the points   A(−3, 4)   B(−1, 4)   C(1, 3)   D(1, 2)
E(−1, 1)   F(1, 0)   F(1, 0)   G(1, −1)   H(−1, −2)
I(−3, −2).

Join the points in alphabetical order and join I to A.

13 Mark the points   A(2, 1)   B(−1, 3)   C(−3, 0)   D(0, −2).

Join the points to make the figure ABCD.
What is the name of the figure?

14 Mark the points   A(1, 3)   B(−1, −1)   C(3, −1).

Join the points to make the figure ABC and describe ABC.

15 Mark the points   A(−2, −1)   B(5, −1)   C(5, 2)   D(2, 2).

Join the points to make the figure ABCD and describe ABCD.

16 Mark the points   A(−3, 0)   B(1, 3)   C(0, −4).

What kind of triangle is ABC?

EXERCISE 16F

Use graph paper and draw your own sets of axes.
Mark a scale on each axis from −8 to +8 using 1 cm for each unit.

In questions **1** to **6** mark the points A and B and then find the length of the line AB.

1 A(2, 2)   B(−4, 2)   **4** A(5, −1)   B(5, 6)

2 A(−2, −2)   B(6, −1)   **5** A(−2, 4)   B(−7, 4)

3 A(4, −4)   B(−4, 2)   **6** A(−1, 2)   B(−8, −2)

In questions **7** to **11** the points A, B and C are three corners of a square ABCD.
Mark the points and find the point D. Give the coordinates of D.

7 A(1, 1)   B(1, −1)   C(−1, −1)

8 A(1, 3)   B(6, 3)   C(6, −2)

9 A(−3, −1)   B(−3, 2)   C(0, 2)

10 A(−2, −1)   B(2, −2)   C(3, 2)

11 A(−3, −2)   B(−5, 2)   C(−1, 4)

In questions **12** to **17** mark the points A and B and the point C, the midpoint of the line AB. Give the coordinates of C.

12 A(2, 2) B(6, 2)   **15** A(2, 1) B(6, 2)

13 A(2, 3) B(2, −5)   **16** A(2, 1) B(−4, 5)

14 A(−1, −2) B(−9, −2)   **17** A(−7, −3) B(5, 3)

MIXED EXERCISE

EXERCISE 16G

Questions **1** to **3** refer to the following diagram.

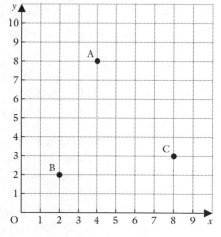

1 Write down

   a the *x*-coordinate of A

   b the *y*-coordinate of B

   c the coordinates of C.

2 Write down the coordinates of a fourth point D such that ABCD is a parallelogram.

3 a Write down the coordinates of the middle point of AC.

   b Write down the coordinates of the middle point of BD.

   c What do you notice?

Questions **4** to **6** refer to the following diagram.

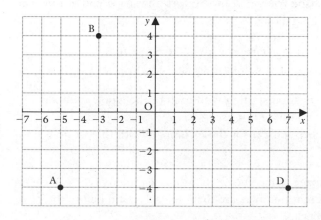

4 Write down

   a the *x*-coordinate of B

   b the *y*-coordinate of D

   c the coordinates of A.

5 a Write down the coordinate of C such that the *x*-coordinates of C is 3 and BC is parallel to AD.

   b What name do you give to the shape ABCD?

6 Find the coordinates of E such that BCDE is a parallelogram. (Notice that the letters should go round the quadrilateral in the order given. BCED is not a parallelogram.)

INVESTIGATION

Study the number chain 2, 8, 32, 11, 5, ...
The next number in the chain is found by multiplying the number of units by 4 and adding the number of tens – if there are any.
For example, the number after 32 is $4 \times 2 + 3 = 11$
and the number after 11 is $4 \times 1 + 1 = 5$

a Write down the next four terms.
What do you notice?

b Write down the first number, then write each of the following numbers in the chain four times. Bracket these numbers off to give a sequence of ordered pairs. The ordered pairs at the beginning of the sequence are

$(2,8)$, $(8,8)$, $(8,32)$, $(32,32)$, $(32,11)$, $(11,11)$, $(11,5)$,

Continue the ordered pairs until you find a good reason to stop!
When you stop state your reason.

c On 2 mm graph paper, draw x- and y-axes and mark a scale on each axis from 0 to 40.
Mark both axes in multiples of 5.
Plot the points that represent the ordered pairs you found in part **b**.
Join each point to the next one with a straight line.
What do you notice when you come to plot the thirteenth pair?

d In part **a** the chain began with 2.
Repeat the investigation starting with 3.
How does the shape you get compare with the shape you got in the previous investigation?

e Repeat the whole thing starting with   **i** 5   **ii** 8
Compare the four shapes you have so far.

f Repeat the investigation starting with each of the whole numbers from 1 to 9 that you have not used so far.
How do the shapes you get for these compare with the previous four shapes?

g Do any of the shapes have line or rotational symmetry?
How many different numbers are there in each chain before the chain starts to repeat?
Compare the sums of the different numbers in each chain.

1 a At Newtown Comprehensive School they always hold important examinations in the school hall. The hall is rectangular and has sufficient space for 8 rows with 16 desks in each row. It is important that each pupil sits in a particular desk.

How can the desks be described so that a particular student can be directed to a particular desk?

Your system could be all numbers, such as (1, 1), (1, 2), ... or all letters, for example Aa, Ab, ... or a mixture of the two, for example A1, A2, ...

Which system is the best for pupils to understand easily and quickly where their place is?

b What system is used when large numbers of people are to be seated, for example in a sports stadium or concert hall?

c How can you remember where your car is parked in an airport car-park where there may be thousands of cars?

d

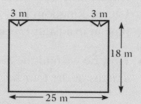

The diagram shows the shape and dimensions of a school hall. Work out a suitable seating plan if the hall is to be used for an examination. The regulations state that the minimum distance between the centres of any two adjacent desks is 1.5 m both from front to back and from side to side. The dimensions of the desks being used are 70 cm by 45 cm. Each desk has an ordinary chair. Don't forget to include an invigilator's desk at a suitable position in the room. Desks can be placed against the walls.

Would the room be more suitable for examinations if the doors were in a different place? Justify your answer.

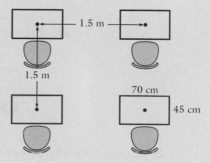

FORMULAS

Suppose you are asked how many biscuits each pupil in a class would get if a packet of biscuits is shared equally among them. You could answer: the number of biscuits each child gets is the number of biscuits in the packet divided by the number of pupils in the class.

The instruction for finding a quantity from other quantities is called a *formula*.

To answer the following questions, instructions or formulas are needed.

- If local time in mainland Europe is 1 hour ahead of local time in the United Kingdom how do I find the time in the UK in terms of the time in mainland Europe?

- How do we find the distance around a rectangular field in terms of the lengths of the four sides?

- How can the Motor Club organiser work out how many boxes of After Eight mints he needs to buy for the annual dinner if he assumes that every person present will eat two?

EXERCISE 17A   In questions **1** to **3** write in words a formula that will enable you to find

1 the number of drawing pins needed to put up posters advertising the school musical if four pins are needed for each poster.

2 the amount of hardcore delivered in a day to a motorway construction site if all the lorries carry the same amount.

3 the total number of legs on all the 4-legged dining chairs in a furniture store.

4 You are told the number of county cricket games taking place each Saturday of the season.
Write a formula in words that will give you the number of

a wickets   **c** bails

b stumps   **d** bats

being used at any one time when all the matches are in progress.
(In a cricket match each of the two wickets has 3 stumps and 2 bails. When play is taking place there are always 2 batsmen.)

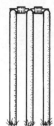

EXPRESSING FORMULAS IN WORDS AND SYMBOLS

There is a rule for changing the numbers in the top row of this table to the numbers in the bottom row.

1	3	5	9	15
7	9	11	15	21

The rule is: the bottom number is equal to the top number plus six.
We know and use symbols for 'is equal to' ($=$) and 'plus' ($+$)
so we can shorten the rule to

$$\text{bottom number} = \text{top number} + 6$$

EXERCISE 17B

In questions **1** to **6** write in words and symbols the rule that tells you how to get the bottom number from the top number.

1

1	2	3	4	5
6	7	8	9	10

2

2	4	6	7	12
4	8	12	14	24

3

5	8	16	21	34
2	5	13	18	31

4

8	12	22	34	46
4	6	11	17	23

5

1	2	3	4	5
4	6	8	10	12

6

2	4	5	7	9	12
4	16	25	49	81	144

Local time in Miami is found by subtracting 5 hours from the time in London.

What time is it in Miami when it is 3 p.m. in London?

Local time in Miami = Time in London − 5 h
= 3 p.m. − 5 h
= 10 a.m.

---

7 The profit made by a company is equal to its income minus its costs. Last year Busicom Computers had an income of £4 m and costs of £3.2 m (£4 m means £4 million, i.e. 4 000 000).

  a Write in words and symbols, a formula to show the connection between income, costs and profit.

  b Use your formula to find the profit made by Busicom last year.

8 The perimeter of a rectangle is found by doubling the length of a longer side and adding it to double the length of a shorter side.

  a Write down a formula connecting the perimeter and the sides.

  b Use your formula to find the perimeter of a rectangle measuring 12 m by 7 m.

9 The capacity of an engine is found by multiplying the capacity of one cylinder by the number of cylinders.
Write down a suitable formula connecting these quantities and use it to find the capacity of an engine that has 4 cylinders, each with a capacity of 498 cubic centimetres.

10 The number of cans of cola on a supermarket shelf is found by counting the number of packs and multiplying by 4.

   a What formula connects the number of cans with the number of packs?

   b At the end of a day an assistant counts 23 packs on the shelves. How many cans are there?

   c She wants to put another 60 cans on the shelves. How many packs is this?

11 First class stamps cost 25 p each and are sold in books of 10. Write a formula that gives

   a the number of stamps in terms of the number of books

   b the cost, in pence, of the stamps in terms of the number of books

   c the cost, in pounds, of the stamps in terms of the number of books

   d the number of books that can be bought in terms of the number of £5 notes paid

   e the number of stamps in terms of the number of £5 notes paid.

12 Answer the questions at the beginning of the chapter.

FORMULAS USING LETTERS FOR UNKNOWN NUMBERS

We are always looking for shorter ways of giving the same amount of information.
On a score sheet you sometimes see o.g. which is short for 'own goal'; we use p to stand for pence, EU for European Union and PO for Post Office.

Things are no different in mathematics: we use a letter for *an unknown number* so that we can write a formula in a shorter way.

To give the number of biscuits each pupil in a class gets when a box containing 60 biscuits is shared out among them, we could use the formula

number of biscuits each pupil gets = 60 ÷ number of pupils in the class

Using n for the number of biscuits each pupil gets
and x for the number of pupils in the class,
we can write the formula as

$$n = 60 \div x \quad \text{or} \quad n = \frac{60}{x}$$

This is a simple example of a formula using letters.

Note that $2 \times w$ or $w \times 2$ can be written $2w$.
To find the value of $2w$ when $w = 5$ we find 2×5,
i.e. when $w = 5$, the value of $2w$ is 10.

EXERCISE 17C

1 Write each of the following formulas in a shorter form.

a $P = 4 \times a$ **c** $w = \frac{3}{4} \times m$

b $Q = p \times 3$ **d** $M = 2 \times y \div 3$

2 Write each of the following formulas in an alternative form
using $\times$ and/or $\div$ signs.

a $X = 5y$ **b** $A = \frac{y}{5}$ **c** $P = \frac{4q}{7}$ **d** $H = \frac{2x}{3}$

In each question from **3** to **7** write down a formula, in letters and
symbols, that connects the two given letters.

3 The number of stones (s) that a person weighs is the number of
pounds (w) the person weighs divided by 14.

4 The number of inches (P) round the edge of a square is four times
the number of inches (x) along the length of one side.

5 The number of kilometres (k) in the distance between two villages,
measured in miles, is the number of miles (m) multiplied by 1.61.

6 The number of eggs (n) in a box is the number of trays (t) in the
box multiplied by 48.

7 The number of components (N) taken out of stock is 10 more than
the number of units to be assembled (n).

Pearl is 8 years older than Carol.

a Write down a formula that expresses Pearl's age (p years) in
terms of Carol's age (q years).

b Use your formula to find Pearl's age when Carol is 30.

a In words, Pearl's age is equal to Carol's age plus 8.
As a formula, using the letters given in the question, this
becomes

$$p = q + 8$$

b If Carol is 30 we replace q by 30 in the formula and so

$$p = 30 + 8 = 38$$

i.e. Pearl is 38 when Carol is 30.

8 Tim is 5 years older than Linda.

a If Tim is T years old and Linda is x years old write down a formula connecting their ages.

b How old is Linda if Tim is 12?

c How old will each person be in 5 years time?

d Will the same formula hold in 5 years time? Justify your answer.

9 A number y is always 13 less than a number x.

a Write down the formula connecting x and y.

b Find y when x is **i** 4 **ii** 16

10 Richard is 28 years younger than his father. Write down a formula for Richard's age (N years) in terms of his father's age (m years).

a Richard's father is 38 now.
How old is Richard?

b How old will Richard be when his father is 60?

11 The charge to send a parcel by special delivery is £2 per kilogram plus a fixed fee of £2.50.
Construct a formula that connects the total cost with the weight of the parcel.
Choose your own letters for this question but remember to give the exact meaning of each one.

12 A father is three times as old as his son.

a If the father's age is H years when the son's age is x years write down a formula connecting H and x.

b Use your formula to find the father's age if the son is 12.

c Will the same formula hold in 12 years time?
If not suggest what the formula will be in 12 years time.

13 One packet of crisps costs 35 p.
Construct a formula that gives the cost of a number of packets of crisps.
Choose your own letters for the unknown numbers and give the meaning of each one.

When a formula is given that connects two related quantities we can find the value of one quantity for any given value of the other. Often the letters that represent the unknown quantities are given. Sometimes we have to choose the letters ourselves.

> The total number of plastic extrusions produced by a machine is found by multiplying the number of hours for which the machine is operating by 15. Choose your own letters to represent the unknown numbers and use the letters to write down a formula connecting them.
>
> Let the total number of extrusions be N when the machine is operating for t hours.
> Total number of extrusions $=$ number of hours $\times$ 15
> $$N = t \times 15$$
> $$N = 15t$$

In this exercise choose your own letters to represent the unknown numbers that can vary. State clearly the meaning you give to each letter. Write down the resulting formula.

1 The profit made by a business is equal to its income minus its costs.

2 The perimeter of a rectangle is found by doubling the length of the longer side and adding to it twice the length of the shorter side.

3 The displacement of an engine is found by multiplying the capacity of one cylinder by the number of cylinders.

4 The number of cans of lemonade delivered to a store is found by multiplying the number of boxes delivered by the number of cans in each box.

> Copy the table
>
> | a | 3 | 8 | 12 | 16 | 30 |
> |-----|---|---|----|----|----|
> | P | | | | | |
>
> and use the formula $P = a + 6$ to complete it.
>
> | a | 3 | 8 | 12 | 16 | 30 |
> |-----|---|---|----|----|----|
> | P | 9 | 14 | 18 | 22 | 36 |
>
> When $a = 3$, $P = 3 + 6 = 9$
> In each case the value of P is 6 more than the value of a.

In questions **1** to **8** copy the table and use the given formula to complete it.

1 $a = b + 7$

b	1	2	3	4	5	6
a						

2 $P = 6a$

Remember that
$6a$ means $6 \times a$

a	1	3	6	8	10	12
P						

3 $M = m - 4$

m	6	8	10	13	17
M					

4 $c = 10 - a$

a	1	2	3	4	5	6
c						

5 $b = 2.5 + c$

c	2	5	5.6	5.8	6.3
b					

6 $p = 12.5 - q$

q	3	5	6.5	7.2	8.9
p					

7 $y = 3.5x$

x	2	4	5	5.6	7.2
y					

8 $p = \dfrac{36}{v}$

v	1	2	3	4	5	8
p						

9 With the formula $t = 20w$ we can work out how long it takes to bake a loaf of bread. w is the number of pounds that the uncooked loaf weighs and t is the number of minutes it takes to bake.

a How many minutes are needed to bake a loaf weighing 2 lb?

b How long should a loaf weighing 3 lb take to bake?

c If a loaf weighs $2\frac{1}{4}$ lb how long will it take to bake?

10 When tyres are bought for a car assembly line the formula $t = 5c$ is used where c is the number of cars to be assembled and t is the number of tyres needed.

   a How many tyres do they need to assemble one car?

   b They plan to assemble 2500 cars next week.
     How many tyres are needed?

   c The last delivery of tyres was an order for 2500.
     How many cars will this supply?

FORMULAS WITH TWO OPERATIONS

A printer quotes a fixed cost of £500 plus £3 per copy to print a quantity of local history books. To work out the cost he uses the formula

$$C = 500 + 3n$$

where C is the number of £s it costs to print n books.
To find the value of C when $n = 2000$, two operations are necessary. First we must multiply 3 by 2000, and secondly we must add the result to 500.

If $n = 2000$, $\quad C = 500 + 3 \times 2000$
$$= 500 + 6000$$
$$= 6500$$

Using this formula, the cost for printing 2000 books is £6500.

EXERCISE 17F

The number of drawing pins (N) needed to pin up q paintings is given by the formula $N = 2q + 2$.
How many pins are needed to pin up 5 paintings?

$$N = 2q + 2$$
If $q = 5$, $\quad N = 2 \times 5 + 2$
$$= 10 + 2$$
$$= 12$$

       ($2q$ is short for $2 \times q$)

$\therefore$ 12 pins are needed to pin up 5 paintings.

Use the formula $N = 2q + 2$, which gives the number of pins needed to hold up q paintings to answer questions **1** to **5**.

1 How many pins are needed to put up 8 paintings?

2 If 13 paintings are put up, how many pins are used?

3 How many pins are used to put up a single painting?

4 If $q = 6$, what is N?

5 What is N when $q = 11$?

Sari is making a matchstick pattern. If she completes n squares she needs M matchsticks where $M = 4n + 1$.
Use this formula to answer questions **6** to **9**.

6 How many matchsticks does she need to complete 6 squares?

7 She completes 9 squares.
How many matchsticks does she use?

8 Find M if $n = 4$.

9 When $n = 12$, what is M?

10 The formula for C in terms of n is $C = 3n + 4$.
Copy the table and complete it.

n	1	4	9	12
C				

11 p and q are connected by the formula $q = 5p - 4$.
Copy the table and complete it.

p	2	5	7	11	15
q					

12 A bicycle manufacturer buys in tyres and uses the formula
$$N = 2a + 10$$
to calculate the number of tyres to order when a is the number of bicycles to be assembled.

a How many tyres should be ordered if 50 bicycles are to be assembled?

b Why do you think 10 has been added to $2a$ in the formula?

FORMULAS WITH TWO SUBSTITUTIONS

In the previous exercise we had to substitute a value for one letter on the right-hand side to find the value of the letter on the left-hand side. In the next exercise two substitutions are required on the right-hand side.

EXERCISE 17G

Use the formula $A = l \times b$, where $A\,cm^2$ is the area of a rectangle of length $l\,cm$ and breadth $b\,cm$, to find the area of a rectangle that is 8 cm long and 5 cm wide.

$$A = l \times b$$
If $l = 8$ and $b = 5$, $\quad A = 8 \times 5$
$$= 40$$
The area of the rectangle is $40\,cm^2$.

Use the formula $A = l \times b$ to find A when

1 $l = 5$ and $b = 3$ **3** $l = 8$ and $b = 3.5$

2 $l = 8$ and $b = 6$ **4** $l = 4.4$ and $b = 2.7$

5 A rectangle is 12 cm long and 6 cm wide.
Find its area.

6 Find the area of a rectangle measuring 20 mm by 15 mm.

If $c = 5a - 2b$, find c when $a = 3$ and $b = 4$.

If $a = 3$ and $b = 4$,
$\quad c = 5 \times 3 - 2 \times 4$
$\qquad = 15 - 8$
$\qquad = 7$

Remember, do multiplication first.

7 If $P = 2l + 2b$ find P when

     a $l = 7.5$ and $b = 2.4$ **b** $l = 5.6$ and $b = 2.8$

8 Given that $I = m \times v$ find I when

     a $m = 2.5$ and $v = 4$ **b** $m = 1.8$ and $v = 6.5$

9 Use the formula $v = u + 10t$ to find v when

     a $u = 3$ and $t = 8$ **b** $u = 8$ and $t = 3$

10 If $p = \dfrac{48}{v}$ find p when

    a $v = 4$ **b** $v = 12$ **c** $v = 16$

11 A printer uses the formula $P = a + 3b$ to work out the cost ($£P$) of printing b books when the set-up costs are $£a$.

    a Can you attach a meaning to the '3' in the formula?

    b Find the cost of printing 1500 books if the set-up costs are £800.

    c Find the value of P when $a = 750$ and $b = 3500$.

DIRECTED NUMBERS

In the last chapter we came across negative numbers for the first time. Sometimes negative numbers are used in formulas, so we need to become more familiar in working with them. Positive and negative numbers together are called *directed numbers*.

Directed numbers can be used to describe any quantity that can be measured above or below a natural zero. For example, a distance of 50 m above sea level and a distance of 50 m below sea level could be written as +50 m and −50 m.

They can also be used to describe time before and after a particular event. For example, 5 seconds before the start of a race and 5 seconds after the start of a race could be written as −5 s and +5 s.

Directed numbers can also be used to describe quantities that involve one of two possible directions. For example, if a car is travelling north at 70 km/h and another car is travelling south at 70 km/h they can be described as going at +70 km/h and −70 km/h.

A familiar use of negative numbers is to describe temperatures. The freezing point of water is 0° centigrade (or Celsius) and a temperature of 5 °C below freezing point is written −5 °C.

Most people would call −5 °C 'minus 5 °C' but we will call it 'negative 5 °C' and there are good reasons for doing so because in mathematics 'minus' often means 'take away'.

A temperature of 5 °C above freezing point is called 'positive 5 °C' and can be written as +5 °C. Most people would just call it 5 °C and write it without the positive symbol.

> A number without any symbol in front of it is a positive number,
>
> i.e. 2 means +2
>
> and +3 can be written as 3

EXERCISE 17H

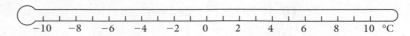

Use the drawing to write the following temperatures as positive or negative numbers.

1 10° above freezing point

4 5° above zero

2 7° below freezing point

5 8° below zero

3 3° below zero

6 freezing point

Write down, in words, the meaning of the following temperatures.

7 −2 °C

10 −10 °C

8 +3 °C

11 +8 °C

9 4 °C

12 0 °C

Which temperature is higher?

13 +8° or +10°

18 −2° or −5°

14 12° or 3°

19 1° or −1°

15 −2° or +4°

20 +3° or −5°

16 −3° or −5°

21 −7° or −10°

17 −8° or 2°

22 −2° or −9°

23 The contour lines on the map below show distances above sea level as positive numbers and distances below sea-level as negative numbers.

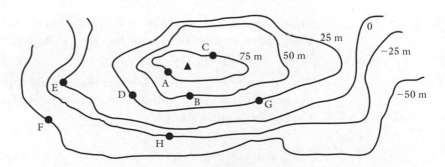

Write down in words the position relative to sea-level of the points A, B, C, D, E, F, G and H.

> Use positive or negative numbers to describe the height of a ball
> thrown up a distance of 5 m.
>
> $+5\,\text{m}$

In questions **24** to **34** use positive or negative numbers to describe the
quantities.

24 5 seconds before blast-off of a rocket.

25 5 seconds after blast-off of a rocket.

26 50 p in your purse.

27 50 p owed.

28 1 minute before the train leaves the station.

29 A win of £50 on premium bonds.

30 A debt of £5.

31 Walking forwards five paces.

32 Walking backwards five paces.

33 The top of a hill which is 200 m above sea-level.

34 A ball thrown down a distance of 5 m.

35 At midnight the temperature was $-2\,°\text{C}$. One hour later it was
1° colder.
What was the temperature then?

36 At midday the temperature was $18\,°\text{C}$. Two hours later it was
3° warmer.
What was the temperature then?

37 A rock-climber started at $+200$ m and came a distance of 50 m
down the rock face.
How far above sea level was he then?

38 At midnight the temperature was $-5\,°\text{C}$. One hour later it was
2° warmer.
What was the temperature then?

39 At the end of the week my financial state could be described as
−25 p. I was later given 50 p.
How could I then describe my financial state?

40 Positive numbers are used to describe a number of paces forwards and
negative numbers are used to describe a number of paces backwards.
Describe where you are in relation to your starting point if you walk
+10 paces followed by −4 paces.

**EXTENDING THE
NUMBER LINE**

If a number line is extended beyond zero, negative numbers can be used
to describe points to the left of zero and positive numbers are used to
describe points to the right of zero.

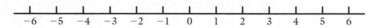

On this number line,   5 is to the *right* of 3
and we say that   5 is *greater* than 3
                or   5 > 3

Also   −2 is to the *right* of −4
and we say that   −2 is *greater* than −4
                or   −2 > −4

So 'greater' means 'higher up the scale'.
(A temperature of −2 °C is higher than a temperature of −4 °C.)

Now   2 is to the *left* of 6
and we say that   2 is *less* than 6
                or   2 < 6

Also   −3 is to the *left* of −1
and we say that   −3 is *less* than −1
                or   −3 < −1

So 'less than' means 'lower down the scale'.

EXERCISE 17I

Draw a number line to help you answer these questions.
In questions **1** to **12** write either > or < between the two numbers.

1 3   2   **5** 1   −2   **9** −3   −9

2 5   1   **6** −4   1   **10** −7   3

3 −1   −4   **7** 3   −2   **11** −1   0

4 −3   −1   **8** 5   −10   **12** 1   −1

In questions **13** to **24** write down the next two numbers in the sequence.

13 4, 6, 8   **17** 9, 6, 3   **21** 36, 6, 1

14 −4, −6, −8   **18** −4, −1, 2   **22** −10, −8, −6

15 4, 2, 0   **19** 5, 1, −3   **23** −1, −2, −4

16 −4, −2, 0   **20** 2, 4, 8   **24** 1, 0, −1

ADDITION AND SUBTRACTION OF POSITIVE NUMBERS

If you were asked to work out $5 - 7$ you would probably say that it cannot be done. But if you were asked to work out where you would be if you walked 5 steps forwards and then 7 steps backwards, you would say that you were two steps behind your starting point.

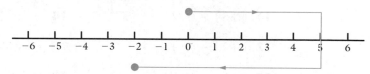

On the number line, $5 - 7$ means

> Start at 0 and go 5 places to the right
> and then go 7 places to the left

So $5 - 7 = -2$

i.e. 'minus' a positive number means   move to the left
and 'plus' a positive number means   move to the right.

In this way $3 + 2 - 8 + 1$ can be shown on the number line as follows:

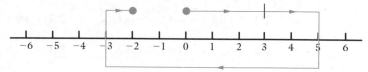

Therefore $3 + 2 - 8 + 1 = -2$.

EXERCISE 17J   Find, using a number line if it helps

1 $3 - 6$ **6** $5 + 2$

2 $5 - 2$ **7** $-2 + 3$

3 $4 - 6$ **8** $-3 + 5$

4 $5 - 7$ **9** $-5 - 7$

5 $4 - 2$ **10** $-3 + 2$

Find $(+4) + (+3)$

$(+4) + (+3) = 4 + 3$
$\qquad\qquad\quad = 7$

Find $(+4) - (+3)$

$(+4) - (+3) = 4 - 3$
$\qquad\qquad\quad = 1$

Find

11 $(+3) + (+2)$ **14** $-(+3) + (+2)$

12 $(+2) - (+4)$ **15** $-(+1) - (+5)$

13 $(+5) - (+7)$ **16** $5 - 2 + 3$

17 $7 - 9 + 4$ **<u>24</u>** $-3 - 4 + 2$

18 $5 - 11 + 3$ **<u>25</u>** $-2 - 3 + 9$

19 $10 - 4 - 9$ **<u>26</u>** $-4 + 2 + 5$

20 $3 + 6 - 10$ **<u>27</u>** $-3 + 1 - 4$

21 $(+3) + (+4) - (+1)$ **<u>28</u>** $(+9) - (+7) - (+2)$

22 $(+2) - (+5) + (+6)$ **<u>29</u>** $-(+3) + (+5) - (+5)$

23 $-(+5) + (+4) - (+8)$ **<u>30</u>** $-(+8) - (+4) + (+7)$

ADDITION AND SUBTRACTION OF NEGATIVE NUMBERS

In a quiz 3 points are awarded for a correct answer and 2 points are deducted for an incorrect answer.

Frank answered the first question correctly so gained 3 points, but he gave incorrect answers to the next three questions.

What was his score after each question?

After 1 question he had 3 points.

After 2 questions the number of points he had was $3 - 2$, i.e. 1.

Now, each wrong answer can be thought of as adding -2 to the score. This means that adding a negative number is the same as subtracting a positive number,

i.e. $3 + (-2) = 3 - 2 = 1$

After 3 questions, the number of points he had was
$1 + (-2) = 1 - 2 = -1$

After 4 questions, his score was $(-1) + (-2) = -1 - 2 = -3$.

Another situation where positive and negative numbers are used regularly is the bank.

Eve is £10 overdrawn.

This could be shown on a statement as -10.

We know that to get the amount she has in the bank back to 0 she must take away the -10, which is another way of saying that she must add 10 to her present balance,

i.e. if, from her balance of -10, she takes away -10 the result is 0.

In a shorter form we can write

$$-10 - (-10) = 0$$

But we know that $\quad -10 + 10 = 0$

So $\qquad\qquad -(-10) = +10$

You can confirm this by using your calculator.

There is a key to change the sign of a number; it is usually $\boxed{\textbf{+/-}}$,

i.e. to get -10 in the display press $\boxed{\textbf{1}}$ $\boxed{\textbf{0}}$ $\boxed{\textbf{+/-}}$

To work out $-10 - (-10)$ key in

$\boxed{\textbf{1}}$ $\boxed{\textbf{0}}$ $\boxed{\textbf{+/-}}$ $\boxed{\textbf{–}}$ $\boxed{\textbf{1}}$ $\boxed{\textbf{0}}$ $\boxed{\textbf{+/-}}$ $\boxed{\textbf{=}}$

The display shows 0.

Find out how to use your calculator to change the sign of a number.

We can now use the following rules:

$$+(+a) = +a \quad \text{and} \quad -(+a) = -a$$
$$+(-a) = -a \quad \text{and} \quad -(-a) = +a$$

i.e. when two signs meet

SIGNS THE SAME MAKE PLUS, SIGNS DIFFERENT MAKE MINUS

EXERCISE 17K

Use a calculator to find

a $5 + (-7)$ **b** $3 - (-7)$ **c** $-4 + (-8)$

a $5 + (-7) = -2$ Key in   5 + 7 +/- =

b $3 - (-7) = 10$ Key in   3 − 7 +/- =

c $-4 + (-8) = -12$ Key in   4 +/- + 8 +/- =

Use a calculator to answer questions **1** to **9**.

1 $3 + (-1)$	**4** $-1 - (-4)$	**7** $5 + (-7)$
2 $5 + (-8)$	**5** $-2 + (-7)$	**8** $-3 - (-9)$
3 $4 - (-3)$	**6** $-2 - (-5)$	**9** $-4 + (-10)$

Find   **a** $3 + (-4)$ **b** $-3 - (-5)$

a $3 + (-4) = 3 - 4 = -1$
b $-3 - (-5) = -3 + 5 = 2$

Do not use a calculator in questions **10** to **21**.

Remember when two signs meet,
signs the same make plus, signs different make minus.

Find

10 $2 - (-8)$	**14** $+2 - (-4)$	**18** $6 - (-3)$
11 $-7 + (-7)$	**15** $-3 + (-3)$	**19** $4 + (+4)$
12 $-3 - (-3)$	**16** $3 + (-2)$	**20** $-5 + (-7)$
13 $+4 + (-4)$	**17** $-3 - (+2)$	**21** $9 - (+2)$

Find $2 + (-1) - (-4)$

$$2 + (-1) - (-4) = 2 - 1 + 4$$
$$= 5$$

Find, without using a calculator

22 $5 + (-1) - (-3)$

30 $12 + (-8) - (-4)$

23 $(-1) + (-1) + (-1)$

31 $9 + (-12) - (-4)$

24 $4 - (-2) + (-4)$

32 $7 + (-3) - (+5)$

25 $-2 - (-2) + (-4)$

33 $2 - (-4) + (-6)$

26 $6 - (-7) + (-8)$

34 $5 + (-2) - (+1)$

27 $9 + (-5) - (-9)$

35 $8 - (-3) + (+5)$

28 $8 - (-7) + (-2)$

36 $7 + (-4) - (-2)$

29 $10 + (-9) + (-7)$

37 $8 - (-2) - (-1)$

38 Use a calculator to check your results to questions **22** to **37**.

Find $-8 - (4 - 7)$

$$-8 - (4 - 7) = -8 - (-3)$$
$$= -8 + 3$$
$$= -5$$

Brackets first

39 $3 - (4 - 3)$

42 $-3 - (7 - 10)$

40 $5 + (7 - 9)$

43 $6 + (8 - 15)$

41 $4 + (8 - 12)$

44 $(3 - 5) + 2$

45 $5 - (6 - 10)$

49 $(7 + 4) - 15$

46 $(4 - 9) - 2$

50 $8 + (3 - 8)$

47 $(3 - 8) - (9 - 4)$

51 $(7 - 12) - (6 - 9)$

48 $(3 - 1) + (5 - 10)$

52 $(4 - 8) - (10 - 15)$

53 Add (+7) to (−5).

54 Subtract 7 from −5.

55 Subtract (−2) from 1.

56 Find the value of '8 take away −10'.

57 Add −5 to +3.

58 Find the sum of −3 and +4.

59 Find the sum of −8 and +10.

60 Subtract positive 8 from negative 7.

61 Find the sum of −3 and −3 and −3.

62 Find the value of twice negative 3.

63 Find the value of four times −2.

64 At the start of the last round in a competition the scores stand at

$$\text{Alf 37, Tanita 24, Colin 35 and Jean 42.}$$

Each competitor gains points for correct answers but loses them for wrong answers.

During the last round Alf gained 24 points and lost 15,
Tanita gained 16 and lost 21,
Colin gained 8 and lost 29,
while Jean gained 26 and lost 14.

Who won the competition?
Who came second, third and last?

65 Newtown is 5 miles from Menton and p miles from Leek. All three villages lie on a stretch of the B3840.
How far is it from Menton to Leek if

a they are on opposite sides of Newton

b they are on the same side of Newton but
  i Leek is furthest away   **ii** Menton is furthest away?

MULTIPLYING AND DIVIDING WITH DIRECTED NUMBERS

We know that $\qquad 2a = a + a$

therefore, if $\quad a = -4, \quad 2a = (-4) + (-4)$

$$= -4 - 4$$

$$= -8$$

But $\qquad 2a = 2 \times a$

so $\qquad 2 \times (-4) = -8$

> A negative number multiplied by a positive number gives a negative number.

We can rewrite $\quad 2 \times (-4) = -8 \quad$ as

$$(-8) = 2 \times (-4)$$

i.e. $\qquad (-8) \div 2 = -4, \quad$ dividing both sides by 2

or $\qquad \dfrac{(-8)}{2} = -4$

> A negative number divided by a positive number gives a negative number.

EXERCISE 17L

Find the value of $\quad$ **a** $5 \times (-6) \qquad$ **b** $(-3) \times 7 \qquad$ **c** $\dfrac{(-14)}{7}$

a $5 \times (-6) = -30$

b $(-3) \times 7 = -21 \qquad$ $\boxed{(-3) \times 7 \text{ is the same as } 7 \times (-3)}$

c $\dfrac{(-14)}{7} = -2$

Find the value of

1 $3 \times (-5)$ $\qquad$ **3** $4 \times (-8)$ $\qquad$ **5** $(-5) \times 4$ $\qquad$ **7** $(-6) \times 6$

2 $6 \times (-8)$ $\qquad$ **4** $(-3) \times 9$ $\qquad$ **6** $(-8) \times 3$ $\qquad$ **8** $4 \times (-6)$

9 $\dfrac{(-6)}{3}$ $\qquad$ **11** $\dfrac{(-24)}{6}$ $\qquad$ **13** $\dfrac{(-26)}{2}$ $\qquad$ **15** $\dfrac{(-36)}{6}$

10 $\dfrac{(-18)}{9}$ $\qquad$ **12** $\dfrac{(-24)}{8}$ $\qquad$ **14** $\dfrac{(-48)}{8}$ $\qquad$ **16** $\dfrac{(-36)}{12}$

FORMULAS THAT INVOLVE DIRECTED NUMBERS

Many formulas involve negative numbers.
One of the most important is the formula

$$c = \tfrac{5}{9}\,(\,f - 32\,)$$

which converts temperature from f degrees Fahrenheit to c degrees Celsius.

EXERCISE 17M

Use the formula $c = \tfrac{5}{9}\,(\,f - 32\,)$ to convert $14\,°F$ to $°C$.

If $f = 14$, $\quad c = \tfrac{5}{9}\,(\,14 - 32\,)$

$$= \tfrac{5}{9} \times (\,-18\,)$$

$$= 5 \times (\,-18\,) \div 9$$

$$= -10$$

$\therefore \quad 14\,°F$ is equivalent to $-10\,°C$

In questions **1** to **8** use the formula $c = \tfrac{5}{9}\,(\,f - 32\,)$ to convert the following temperatures into $°C$.

1 $5\,°F$ **3** $32\,°F$ **5** $-13\,°F$ **7** $10.4\,°F$

2 $23\,°F$ **4** $-4\,°F$ **6** $24.8\,°F$ **8** $-12.5\,°F$

9 If $v = u + 5t$ find v when

  a $u = -10$ and $t = 1$ **b** $u = -15$ and $t = 0.5$

10 Given that $P = q + 2r$, find P when

  a $q = -4$ and $r = 3$ **c** $q = -3$ and $r = -2$

  b $q = 6$ and $r = -4$ **d** $q = 8$ and $r = -4$

11 Given that $S = 2a + 3b$ find S when

  a $a = 4$ and $b = -3$ **b** $a = -3$ and $b = 5$

12 If $y = 3x + c$ find y when

  a $x = 6$ and $c = -5$ **c** $x = -3$ and $c = -1$

  b $x = -4$ and $c = 7$ **d** $x = -7$ and $c = 9$

13 If $y = 5x - c$ find y when

  a $x = 4$ and $c = -5$ **c** $x = -3$ and $c = -9$

  b $x = -3$ and $c = 5$ **d** $x = 2$ and $c = -10$

14 Given that $A = 5b - 2c$ find A when

  a $b = -3$ and $c = 9$ **c** $b = -7$ and $c = 4$

  b $b = 2$ and $c = -6$ **d** $b = -0.05$ and $c = -1.5$

15 Given that $P = 3q - 2r$ find P when

  a $q = 0.7$ and $r = -3.5$ **b** $q = -0.55$ and $r = 3.4$

**MIXED
EXERCISES**

EXERCISE 17N

1 Each pack delivered to an electronics company by Modcomp Ltd contains 48 components.

  a Write down a formula that connects the total number (N) of components delivered with the number (n) of packs delivered.

  b Use your formula to find the number of components in 12 packs.

2 The formula connecting p, q and r is $r = 5p - 2q$.
Find the value of r when

  a $p = 4$ and $q = 3$ **b** $p = 4.7$ and $q = 1.3$

3 Which is the higher temperature, $-5°$ or $-8°$?

4 Write $<$ or $>$ between

  a -3 2 **b** -2 -4

5 Find   **a** $-4 + 6$ **b** $3 + 2 - 10$ **c** $2 + (-4)$

6 Find   **a** $3 - (-1)$ **b** $-2 + (-3) - (-5)$ **c** $4 - (2 - 3)$

7 If $y = 3x + c$ find the value of y when

  a $x = -2$ and $c = 8$ **b** $x = 4$ and $c = -7$

EXERCISE 17P

1 60 W electric light bulbs are sold in packs of six.

    a Write down a formula that gives the total number of bulbs (N) in terms of the number of packs bought (n).

    b Anne buys 6 packs. How many bulbs has she bought?

2 Given that $E = Ri$ find E when

    a $R = 8$ and $i = 2$ **b** $R = 4.5$ and $i = 0.6$

    Note that Ri means $R \times i$.

3 Which is the lower temperature, $0°$ or $-3°$?

4 Write $<$ or $>$ between

    a 3 -4 **b** -7 -10

5 Find   **a** $2 - 8$ **b** $3 - 9 + 4$ **c** $(+2) - (-3)$

6 Find   **a** $(-4) - (-5)$ **b** $3 + (5 - 8)$ **c** $-2 - (4 - 9)$

7 Given that $P = 12 - 5q$ find P when

    a $q = 5$ **b** $q = -6$

INVESTIGATIONS

1 Try this on a group of pupils or friends.

Think of a number between 1 and 10.
Add 4.
Multiply the result by 5.
Double your answer.
Divide the result by 10.
Take away the number you first thought of.
Write down your answer.

However many times you try this the answer is 4.
Investigate what happens when you use numbers other than whole numbers between 1 and 10. Try, for example, larger whole numbers, decimals, negative whole numbers, fractions.
Is the answer always 4?
Try to use the algebra you have learned in this chapter to explain your answer.

2 a

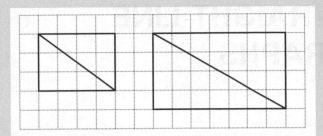

On squared paper mark out several rectangles of different sizes. The length (l) and breadth (b) of each rectangle must be a whole number of squares with a common factor of 1 only. For example, sides 6 squares by 7 squares is acceptable, but 4 squares by 6 squares is not because 4 and 6 have a common factor 2.

For each rectangle draw a diagonal and count the number of squares (d) through which the diagonal passes. Two examples are given above.

Gather all your results together in a table like the one given below. Now find a formula that connects d with l and b.

You should collect the data for at least 6 rectangles before you try to write your formula. When you think you have found the formula, test it on some other rectangles.

l	b	d
4	3	6
7	4	10

b When you are satisfied that you have found the correct formula draw rectangles that have sides with a highest common factor greater than 1. For example, a rectangle 6 squares by 9 squares will do, or one 8 squares by 12 squares.

You will find it helpful to use the following table. The letters l, b and d have the same meaning as in part **a**. h is the highest common factor of l and b. For example if $l = 6$ and $b = 4$ then $h = 2$

l	b	h	d
6	4	2	8

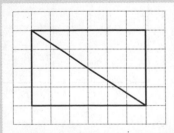

Find a formula that connects l, b, h, d.

STRAIGHT LINE GRAPHS

18

A graph is a diagram showing the relationship between two varying quantities.

LINE GRAPHS

When Joyce went into hospital her temperature was taken every hour and recorded on this chart. The points representing the temperatures were joined by straight lines. A graph like this is called a *line graph*. The result gives a clear picture of what is happening to her temperature as the hours pass.

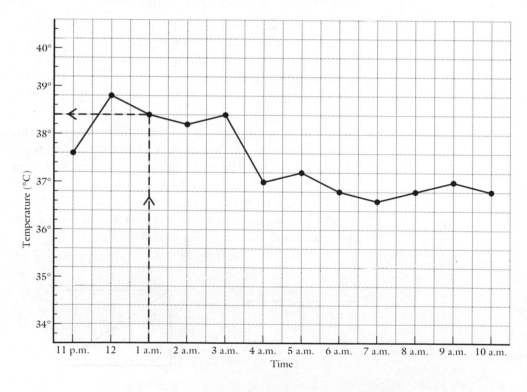

To find Joyce's temperature at 1 a.m., first find 1 a.m. on the time scale, then go up to the graph and straight across to read the value on the temperature scale. The value is **38.4** °C.

EXERCISE 18A

1 Use the graph opposite to answer these questions.

a What was Joyce's highest recorded temperature and when did it occur?

b What was her lowest recorded temperature?

c By 6 a.m. her temperature had returned to normal. What is her normal temperature?

d How long was she in hospital before her temperature returned to normal?

e Can you say what her highest temperature was while she was in hospital?
Give a reason for your answer

2

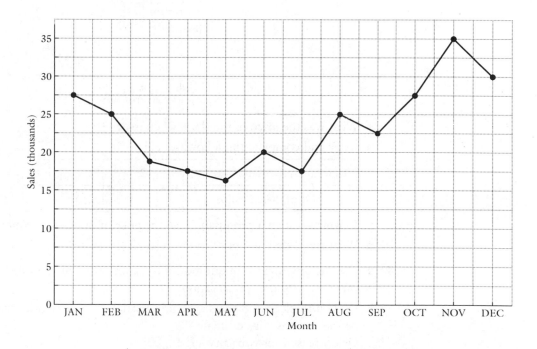

The graph shows the total monthly sales at Benshaw plc.

a In which month were the sales

i greatest **ii** least?

b Does the graph suggest gradually increasing sales?

c Does it look as though the sales improve around Christmas?

d Assuming that the sales graph has looked like this every year for the last five, can you think of a business that could have this sales pattern?

3 The line graph shows the price of a share in a privatised company at yearly intervals after privatisation.

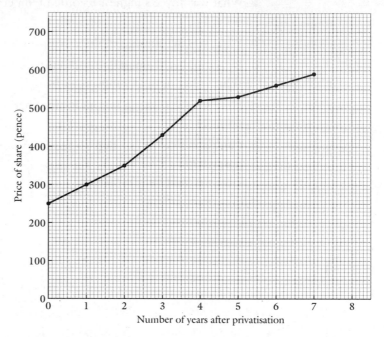

a What was the price of the share when issued?

b Copy and complete the following table.

Time after privatisation (years)	1	2	3	4	5	6
Value of share (pence)						

c In which year after privatisation did the price of the share rise most?

d What trend do you notice in the price of the share?

4 **a** Use the graph given in question **3** to estimate the price of the share, 8 years after privatisation.

b Do you think that, 8 years after privatisation, the share price will be the same as your estimate?
Give a reason for your answer.

5 Use the graph for question **2** for these questions.

a Can the graph be used to find the half-monthly sales figures for the year? Explain your answer.

b Looking at this graph, the managing director asked why the sales had fallen in the first half of July. The sales director replied that they had not; they had in fact increased for the first two weeks of July. How could the sales director justify this statement and, assuming it is correct, describe what happened to the sales in the second half of July.

**CONVERSION
GRAPHS**

In a line graph, the lines between the points on the graphs show the general shape only. We cannot find values between points.

For example, in the temperature chart at the beginning of this chapter, Joyce's temperature was taken at 3 a.m. and at 4 a.m. It was not taken between these times so we have no way of knowing what Joyce's temperature was at, say, 3.30 a.m; We certainly cannot use the line between the points on the graph to find this.

In the graphs that follow, the lines have meaning along the whole of their length.

EXERCISE 18B

1 Conversion graphs are very useful for changing from one system of units to another. This graph can be used to convert between pints and litres.

For example, to find the rough equivalent of 15 pints in litres, we find 15 on the horizontal (pints) axis, go up to the line and then across to the vertical (litres) axis. The reading here is about 9 litres,

i.e. 15 pints ≈ 9 litres.

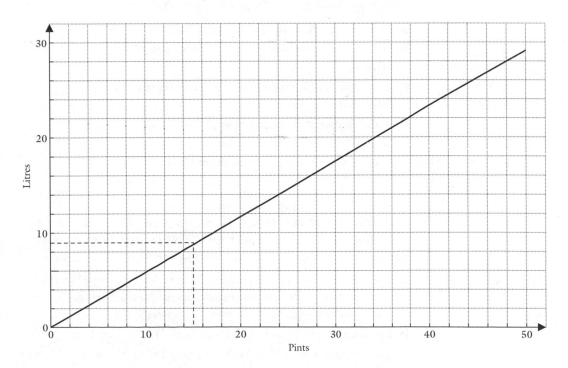

a Use the conversion graph to find the rough equivalent in litres of
i 10 pints   **ii** 24 pints   **iii** 46 pints
b Find the rough equivalent in pints of
i 10 litres   **ii** 24 litres   **iii** 13 litres

2

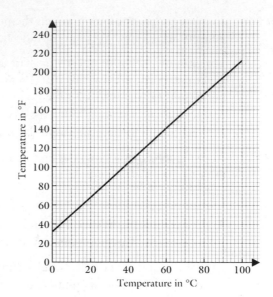

This graph converts between temperature in degrees Celsius and in degrees Fahrenheit. Use the graph to answer the questions that follow.

a Convert **i** 40 °C to °F   **ii** 140 °F to °C

b On a warm sunny day the temperature was 25 °C.
What is this in °F?

c On a cold day last winter the temperature was 40 °F.
What is this in °C?

d Water freezes at 0 °C and boils at 100 °C.
Give these temperatures in °F.

e Normal body temperature is about 98 °F.
What is this in °C?

3 The freezing point of water is 0 °C, but temperatures can fall below this value. We can describe a temperature below freezing using negative numbers, for example 2 °C below freezing can be written −2 °C.

a Copy the graph used in question **2** on to squared paper so that there is space to the left and below. Extend the axes to the left and down, and extend the sloping line backwards.

b Use your graph to convert **i** −2 °C to °F   **ii** 5 °F to °C

c The temperature in New York on a cold February day in 1992 was −2 °F. What is this in °C?

d The temperature in Moscow on a very cold day last winter was −10 °C. What was this in °F?

e When an explorer reached the north pole the temperature was −35 °C. Convert this temperature into °F.

COORDINATES
AND STRAIGHT
LINES

EXERCISE 18C

1 The points A, B, C, D and E are all on the same straight line.

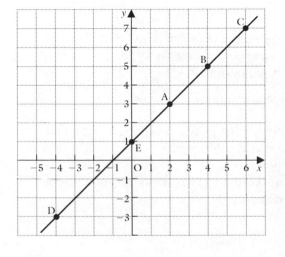

a Copy and complete this table for the coordinates of the points A, B, C, D and E.

	A	B	C	D	E
x					
y					

b F is another point on the same line.
The x-coordinate of F is 5.
Write down the y-coordinate of F, and add F to the table.

c What is the connection between the y-coordinate and the x-coordinate of points on this line?

d G, H, I, J, K, L and M are also points on this line.
Fill in the missing coordinates.

$G(8, \square)$ $H(10, \square)$ $I(-4, \square)$ $J(\square, 12)$
$K(\square, 18)$ $L(\square, -10)$ $M(a, \square)$

2 The points A, B, C, D and E are all on the same straight line.

a Write down the coordinates of the points A, B, C, D, E, placing them in a table like the one given for question **1**.

b H is another point on this line.
Its x-coordinate is 8;
what is its y-coordinate?

c How is the y-coordinate of each point related to its x-coordinate?

d I, J, K, L, M, and N are further points on this line.
Fill in the missing coordinates.

$I(12, \square)$ $J(20, \square)$ $K(30, \square)$
$L(-12, \square)$ $M(\square, 9)$ $N(a, \square)$.

3

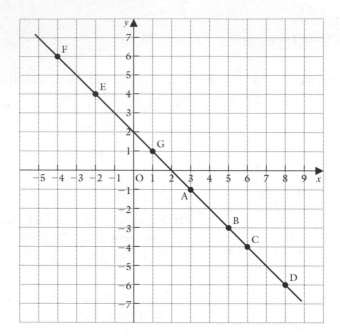

The points A, B, C, D, E, F and G are all on the same straight line.

a Write down the coordinates of the points A, B, C, D, E, F and G.

b H, I, J, K, L, M, N, P and Q are further points on the same line.
Fill in the missing coordinates.

H$(7, \square)$ I$(10, \square)$ J$(12, \square)$ K$(20, \square)$ L$(-7, \square)$
M$(-9, \square)$ N$(\square, 10)$ P$(\square, -8)$ Q$(\square, 12)$

c How is the y-coordinate related to the x-coordinate?

PRACTICAL
WORK

1 Collect examples of graphs from newspapers and magazines or from school textbooks in other subjects.

a Which of your graphs are like questions **1** and **2** in **EXERCISE 18A**, i.e. graphs where the lines have little or no meaning between the points that they join?

b Which graphs have meaning along the whole of their lengths?

2 What data could you collect that gives a straight line graph that has meaning along the whole of its length?
Gather such data and plot it on a graph.
Explain how you can use your graph to give information that was not available to you before you drew the graph.

SUMMARY 4

PERIMETER

The perimeter is the distance all round the edge of a shape.

AREA

Area is measured in standard sized squares.
A square with side 1 cm long is called one square centimetre and written 1 cm^2.

$1\,\text{cm}^2 = 10 \times 10\,\text{mm}^2 = 100\,\text{mm}^2$
$1\,\text{m}^2 = 100 \times 100\,\text{cm}^2 = 10\,000\,\text{cm}^2$
$1\,\text{km}^2 = 1000 \times 1000\,\text{m}^2 = 1\,000\,000\,\text{m}^2$

The area of a square $= (\text{length of a side})^2$

The area of a rectangle $= \text{length} \times \text{breadth}$

SPECIAL QUADRILATERALS

In a square

- all four sides are the same length
- both pairs of opposite sides are parallel
- all four angles are right angles.

In a rectangle

- both pairs of opposite sides are the same length
- both pairs of opposite sides are parallel
- all four angles are right angles.

In a rhombus

- all four sides are the same length
- both pairs of opposite sides are parallel
- the opposite angles are equal.

In a parallelogram

- the opposite sides are the same length
- the opposite sides are parallel
- the opposite angles are equal.

In a trapezium

- just one pair of opposite sides are parallel.

PARALLEL LINES   When two parallel lines are cut by a transversal

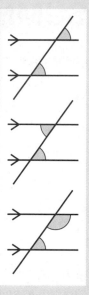

- the corresponding angles are equal

- the alternate angles are equal

- the interior angles add up to 180°.

COORDINATES   Coordinates give the position of a point as an ordered pair of numbers, e.g. (2, 4).
The first number gives the distance from O in the direction of the x-axis, and is called the x-coordinate.
The second number gives the distance from O in the direction of the y-axis, and is called the y-coordinate.

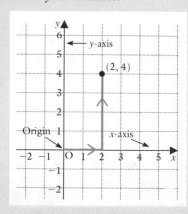

DIRECTED NUMBERS   Positive and negative numbers are collectively known as directed numbers. They can be represented on a number line.

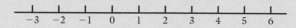

The rules for adding and subtracting directed numbers are

$$+(\,+a\,) = a \qquad\qquad +(\,-a\,) = -a$$
$$-(\,+a\,) = -a \qquad\qquad -(\,-a\,) = a$$

FORMULAS

A formula is a general rule for finding one quantity in terms of other quantities, e.g. the formula for finding the area of a rectangle is given by

$$\text{Area} = \text{length} \times \text{breadth}$$

When letters are used for unknown numbers, the formula can be written more concisely, i.e. the area, A cm², of a rectangle measuring l cm by b cm, is given by the formula

$$A = l \times b$$

REVISION EXERCISE 4.1 (Chapters 14 & 15)

1

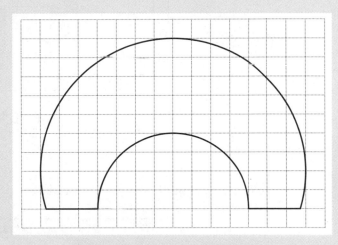

Each square represents 1 cm².
By counting squares find, approximately, the area of the shape.

2 Find the area of each shape.

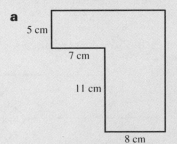

a   5 cm   7 cm   11 cm   8 cm

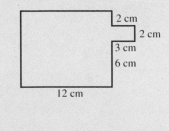

b   2 cm   2 cm   3 cm   6 cm   12 cm

3 The table gives some measurements of two rectangles.
Copy the table and fill in the missing values.

Rectangle	Length	Breadth	Perimeter	Area
A	6 cm		18 cm	
B		5 cm		50 cm²

4 Change   **a** 5.6 cm² into mm²   **b** 5 000 000 m² into km²

5 A rectangular carpet measures 3 m by 4.5 m.
Find   **a** its area   **b** the cost of cleaning it at 80 p per square metre.

6

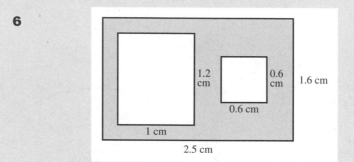

Find the area of the cork (shown shaded) in this gasket.

7 Find the size of the angle marked *d*.

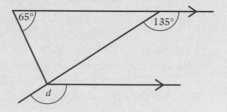

8 Find the size of each of the marked angles.

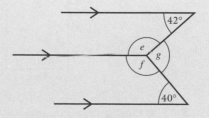

9 Copy these statements and fill in the blanks.

a The _____ of the three angles of a triangle is 180°.
b Corresponding angles are _____ .
c 56° and _____ are supplementary angles.

10

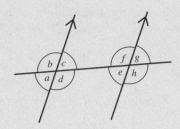

Use the letters marked in the diagram to give

a two angles that are corresponding angles, one of which is *d*
b two angles that are interior angles, one of which is *e*
c two angles that are supplementary angles, one of which is *d*
d two angles that are alternate angles, one of which is *e*

**REVISION
EXERCISE 4.2
(Chapters 16 to 18)**

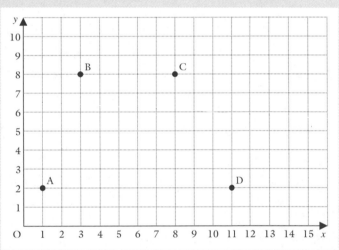

Copy this diagram and use your copy to answer questions **1** to **4**.

1 Write down the coordinates of **a** A   **b** D

2 Join the points ABCDA in order.
What name do you give to this shape?

3 Mark a point E so that ABED is a parallelogram.
Write down the coordinates of E.

4 Write down the coordinates of M, the middle point of AD.
Join MC.
How is CM related to ED?

5 Which is the higher temperature, $-3\,°C$ or $-5\,°C$?

6 The formula connecting p, q and r is $r = 3p - 2q$.
Find the value of r when

    a $p = 4$ and $q = 5$ **b** $p = -3$ and $q = 4$

7 Subtract 4 from the sum of -5 and 8.

Questions **8** to **10** refer to the diagram given below.

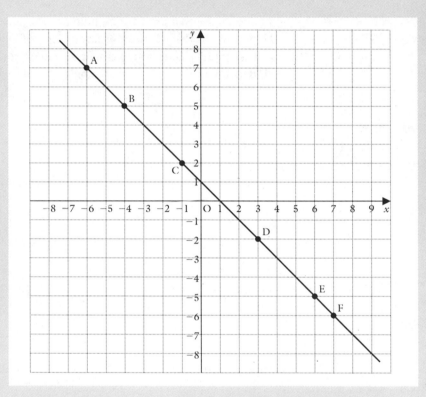

8 The points A, B, C, D, E and F are all on the same straight line.
Write down the coordinates of each point.

9 G, H, I, J, K and L are further points on the line.
Fill in the missing coordinates.
G$(-2,\ \ )$, H$(5,\ \ )$, I$(\ \ ,6)$ J$(\ \ ,-1)$, K$(-5,\ \ )$,
L$(\ \ ,-7)$.

10 Use your answers to questions **8** and **9** to decide how the
x-coordinate is related to the y-coordinate for any point that lies on
the given line.

**REVISION
EXERCISE 4.3
(Chapters 14 to 18)**

1 Change **a** $4400 \, \text{cm}^2$ into m^2 **b** $790\,000 \, \text{mm}^2$ into m^2

2 Five squares, each of side 4 cm, are
arranged to form a cross as shown
in the diagram.

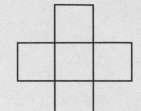

  a Find **i** its perimeter **ii** its area
  b Suppose the squares are
  rearranged in a row.

How does the length of the perimeter
compare with the perimeter of the cross in part **a** ?

3 Find the value of three times negative three.

4 Write $<$ or $>$ between each pair of numbers.

  a -4 6 **b** -7 -3 **c** -5 -9

5 The temperature when I got up at 6 a.m. yesterday was $-4\,°\text{C}$. By
noon it had increased by $6\,°\text{C}$. From noon to 6 p.m. the
temperature dropped by $10\,°\text{C}$.

  a What was the temperature at 6 p.m. ?

  b How much did the temperature fall between 6 p.m. and
  midnight if the temperature at midnight was $-11\,°\text{C}$?

6 Use the formula $f = \dfrac{9c}{5} + 32$ to find the value of f when

  a $c = 40$ **b** $c = 0$ **c** $c = -10$

Draw your own set of axes and mark a scale on each one from -10 to 10.
Use them to answer questions **7** and **8**.

7 Plot the points $P(-8, -3)$, $Q(-4, 2)$, $R(0, -3)$ and
$S(-4, -8)$.
What name do you give to the shape PQRS ?

8 a Write down the coordinates of T if its y-coordinate is the same as
  the y-coordinate of Q and its x-coordinate is minus the
  x-coordinate of Q.

  b Find the coordinates of U if its x-coordinate is the same as the
  x-coordinate of T and its y-coordinate is the same as the
  y-coordinate of R.
  What name do we give to the shape QTUR ?

9 Find each angle marked with a letter.

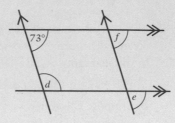

10

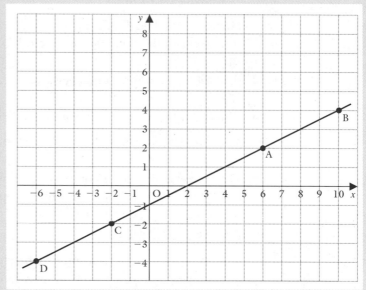

The points A, B, C and D are all on the same straight line.

a Write down the coordinates of these four points.

b F is another point on this line. The x-coordinate of F is 4. Write down the y-coordinate of F.

c G is another point on this line. The y-coordinate of G is −1. What is the x-coordinate of G?

REVISION EXERCISE 4.4
(Chapters 1 to 18)

1 Subtract one thousand eight hundred and seventy-nine from three thousand two hundred and forty-eight.

2 Find the value of   **a** 3^4 **b** 7^3 **c** $2^2 \times 5^3$

3 **a** Estimate the value of $647 \div 19$

    b Use a calculator to find $647 \div 19$ correct to 2 decimal places.

    c Find $36 \div 9 - 2 + 15 \times 7 \div 3$

4 **a** Which is the larger: $\frac{8}{9}$ or $\frac{9}{10}$?

    b Which is longer: a pipe 322 mm long or one 300 cm long?

5 Find, giving your answer in the unit in brackets

   a $4.5\,\text{m} + 85\,\text{cm}$ (cm)   **b** $3\,\text{m} + 165\,\text{cm} + 40\,\text{mm}$ (cm)

6 At the beginning of term a teacher has 12 packs of exercise books, each pack containing 20 books.

   a How many exercise books does she have altogether?

   b She has 27 pupils in her form.
      What is the largest number of exercise books she can give out so that every pupil has the same number of books?

7 Construct a triangle ABC in which the length of the side AB is 9.7 cm, the length of the side AC is 10.6 cm and the length of the side BC is 8.3 cm.

8 Which is heavier, a 5 lb bag of potatoes or a 2 kg bag of sugar?

9 One letter is chosen at random from the letters in the word EFFORT. What is the probability that the letter is

   a R   **b** F   **c** a consonant?

10 Use the formula $R = p \times q$ to find the value of R when

   a $p = 3.5$ and $q = 3$ **c** $p = 4$ and $q = -1.5$

   b $p = 2.4$ and $q = 5.5$ **d** $p = -3$ and $q = -2.5$

**REVISION
EXERCISE 4.5
(Chapters 1 to 18)**

1 In an election there were three candidates. 1273 voted for Bullen, 942 for Wilson and 372 for Chan.

   a How many people voted?

   b If 7261 people were entitled to vote, how many failed to do so?

2 **a** Find $18 \div 6 \times 3 + 15 \times 2 - 42 \div 7$

   b Find **i** $2\frac{7}{13} + 1\frac{1}{2}$ **ii** $11\frac{4}{7} - 5\frac{1}{3}$

   c Express $\frac{13}{20}$ as **i** a decimal   **ii** a percentage

3 Find, without using a calculator

   a $0.42 + 1.37 - 2.45$ **c** $0.059 \div 100$

   b 0.043×10 **d** $5.727 - 2.9$

4 **a** Estimate the value of $169.4 \div 52$, and then use a calculator to find its value correct to 2 decimal places.

   b Express 0.48 as **i** a percentage **ii** a fraction in its lowest terms

   c Express $\frac{4}{13}$ as a decimal correct to 2 decimal places.

5 Find the value of the angles marked with a letter.

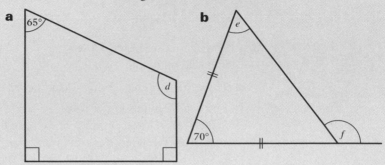

6 Arrange the following lengths in order of size with the smallest first.

0.75 m,   693 mm,   73 cm,   2.4 m,   610 cm

7 The diagram shows the floor plan of Sean's bedroom.

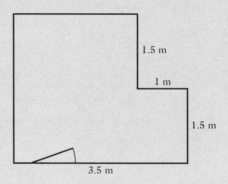

Find

a the length of skirting board used if the door is 80 cm wide

b the area of the floor in square metres.

8 The coordinates of three corners of a square ABCD are
A($-3, -4$), B($-2, 3$), C($5, 2$).
Find the coordinates of D.

9 a Write down two angles that are supplementary.

b Draw a diagram to show an example of
i corresponding angles
ii alternate angles
iii vertically opposite angles

10 Simplify

a $4 + (7 - 5)$ **b** $9 - (7 - 4)$ **c** $(5 - 7) + (2 - 7)$

SUMMARISING AND COMPARING DATA

19

SUMMARISING
INFORMATION

Aisha would like to have a personal cassette player and decides to save up to buy one. However she has no idea about the likely cost, so she needs some information. A detailed price list of all available models can be confusing and some sort of summary of the prices could be more useful. For instance, it might help Aisha if she is told that

- prices vary from about £8 to £120
- average price is £35
- a middle-of-the-range one costs about £30
- the most common cost is around £20.

EXERCISE 19A

In each of the following situations, discuss whether it would be more useful to you to have detailed figures or some form of summary.

1 You are going on holiday and want to know what weather to expect.

2 You need to buy some soft drinks for an end-of-term party for your class.

3 You want to know if there are more people taller than you than there are shorter than you.

4 You want to know how much you would have to pay for a new multimedia computer.

**RANGE AND
CENTRAL
TENDENCY**

We often need to represent a set of numbers by one representative number which gives an indication of the middle of the set. It is also useful to have an indication of the spread of the numbers.
The first piece of information about the cost of personal cassette players gives an idea of the spread of prices.

We can give an indication of the spread of a list of values by giving their *range*.

> The range is the difference between the largest and smallest values.

The range of prices for the cassette players is £120 − £8 = £112.

Each of the last three pieces of information about the cost of a personal cassette player illustrates one way of indicating a central price.

Figures which give an indication of the middle of a set are called *measures of central tendency*. There are three of these and we will look at each one in turn.

MEAN VALUE

The most commonly used central figure is the arithmetic average or *mean*.
This is the figure that results from sharing out the different values equally.

For example,

Jan   Raj   Sara

Jan, Raj and Sara searched for some pebbles to use for skimming on water.
Jan found 9 pebbles, Raj found 4 and Sara found 5 pebbles.
They decided to put the pebbles into one pile and divide them up equally.

There are 18 pebbles in total to be shared among the three of them, so this gives 6 pebbles each.
The 6 pebbles that each child then has is called the *mean* of the 9, 4 and 5 pebbles that the children found.

To find the mean we add up all the values and divide by the number of values

$$\text{Mean} = \frac{\text{Sum of all the values}}{\text{Number of values}}$$

The mean is not always a whole number, or even a quantity that can exist.

For example, if Jan has 2 dogs, Raj has 1 dog and Sara has 1 dog, then this gives 4 dogs in total.
If they could be shared equally each of the three children would have $\frac{4}{3} = 1\frac{1}{3}$ dogs, and this is clearly impossible.

But $1\frac{1}{3}$ *is* the mean of the *numbers* 2, 1 and 1.

> In five tests, Alan received marks of 7, 8, 7, 9 and 4.
>
> **a** What is his mean mark?
> **b** What is the range of his marks?
>
> **a** Total marks scored $= 7 + 8 + 7 + 9 + 4$
> $\qquad\qquad\qquad = 35$
> There are 5 marks.
> Mean mark $= 35 \div 5$
> $\qquad\qquad = 7$
>
> **b** The highest mark is 9 and the lowest mark is 4.
> The range of the marks is $9 - 4 = 5$

1 Six pupils got the following marks in a test: 5, 7, 8, 1, 8, 1.
What is the mean mark?

2 In three different shops, the price of a can of cola is 27 p, 25 p and 23 p.
a What is the mean price?
b What is the range of the prices?

3 Five people decided to pool their money. They put in the following amounts:
£10, £5, £6, £7 and £12.
a How much was in the pool?
b If the five people had contributed equally to this total, how much would each have given?
c What was the mean amount contributed to the pool?

4 The ages of the children in a swimming club are
9, 10, 8, 10, 11, 8, 12, 9, 10, 11, 10, 12
Find the mean age and the range of ages.

5 Find the mean and range of
a 2, 4, 8, 4, 7, 1, 7, 6, 5, 6
b 12, 15, 13, 10, 24, 16
c 24, 35, 44, 28, 34
d 1.2, 1.5, 1.3, 1.2
e 12.4, 16.5, 27.9, 3.5, 26.1

6 The buses that passed the school gate in four hours were counted.
From this information it was found that the average number of buses per hour was 3.
How many buses were counted?

7 The 28 pupils in Class 7R took a maths test. The average mark was 15. Carlos was away on the day of the test and took it later. His mark was 24.

   a Will Carlos's mark increase or decrease the average mark for the test?

   b What is the total of the marks for all the children who took the test at the proper time?

   c Find the new mean mark when Carlos's mark is added into the total.

FINDING THE MEAN FROM A FREQUENCY TABLE

This table shows the marks in a maths test.

Mark	Frequency
0	1
1	1
2	8
3	11
4	5
5	4
	Total: 30

From the figures in the table we can guess that the mean mark is about 3. To find the mean mark we first need to add up all the marks.

We could do this by listing each mark, but it is quicker to find the sum of the 0s, 1s, 2s, ... separately and then add these up. We can do this easily direct from the table.

For example, there are eight 2s, so the 2s add up to $8 \times 2 = 16$.

We add another column to the table so that we can keep track of what we are doing.

Mark	Frequency	Frequency $\times$ Mark
0	1	0
1	1	1
2	8	16
3	11	33
4	5	20
5	4	20
	Total: 30	Total: 90

Now we can see that there are 30 pupils and their marks add up to 90, so the mean mark is $90 \div 30 = 3$
(This agrees with our guess.)

EXERCISE 19C

1 The pupils in Class 7G gathered this information about themselves.

  a Guess the mean number of children per family.

  b Find the mean number of children per family.

Number of children in each family	Frequency
1	8
2	12
3	4
4	2

2 Joshua tossed three coins several times and recorded the number of heads that showed at each toss. His results are shown in the table.

  a Guess the value of the mean.

  b Find the mean number of heads per toss.

Number of heads obtained when three coins are tossed	Frequency
0	9
1	7
2	16
3	3

  c Is it better to give the answer for the mean as a fraction or as a decimal correct to 1 decimal place? Give reasons for your answer.

3 Once every five minutes, Debbie counted the number of people queuing at a checkout and gave her results in this table.

  a How many times did she count?

  b What is the mean number of people queuing?

Number of people queuing at a supermarket checkout	Frequency
0	4
1	6
2	5
3	2
4	2

4 The children in Class 7P were asked to count the number of one pound coins that they had with them. The distribution of these coins is shown in the bar chart.

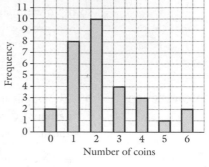

a How many children took part?

b What is the total value of the one pound coins?

c What is the range of the number of one pound coins?

d If the total sum of money represented here was shared out equally among the children, how much would each child have?

e What is the mean number of one pound coins?

USING MEAN AND RANGE TO COMPARE TWO DISTRIBUTIONS

So far in this cricket season, Tom Batt has played five innings. His scores were 22, 53, 40, 35 and 25 so his mean score is 35. Reg Wicketaker has also completed five innings of 26, 90, 0, 52 and 17 so his mean score is 37.

There is little difference between the mean scores but the range of Tom's scores is $53 - 22 = 31$ and the range of Reg's scores is $90 - 0 = 90$.

Comparing the two batsmen's scores indicates that, although Tom Batt has a slightly lower batting average, his scores are the more consistent of the two.

EXERCISE 19D

1 In the end-of-term tests, nine subjects were set and each one was marked out of 20. Sandra took eight subjects and her marks were

12, 16, 14, 9, 8, 20, 15 and 10.

Karen took only five subjects and scored

10, 15, 11, 14 and 10.

a On average, which girl did better?

b Which girl was more consistent in the standard she achieved?

2 Mr and Mrs Burton each made a batch of raisin cookies for a stall at the school fête. Out of curiosity they weighed each cookie and found that Mr Burton's weighed

20, 25, 16, 21, 24, 26, 13, 17, 22 and 16 grams.

Mrs Burton's weighed

22, 21, 18, 17, 20, 20, 21, 19, 20 and 22 grams.

Compare the means and ranges of the weights of the two batches and comment on them.

3 These two bar charts illustrate the results of the same test given to Group 7P and Group 7B.

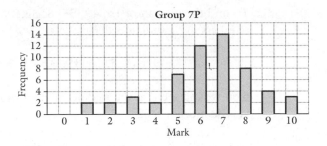

Group 7P

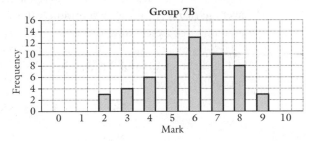

Group 7B

a Find the mean and range of each set of marks.

b Compare the two sets of marks.

MEDIAN

The median value of a set of numbers is the value of the middle number when they have been placed in ascending (or descending) order of size.

Imagine nine children arranged in order of their height.

Median value 154 cm

↑
Middle child

The height of the fifth or middle child is 154 cm,
i.e. the median height is 154 cm.

Similarly 24 is the median of 12, 18, 24, 37 and 46; two numbers are smaller than 24 and two are larger.

To find the median of 16, 49, 53, 8, 32, 19 and 62, first rearrange the numbers in ascending order:

$$8, 16, 19, 32, 49, 53, 62$$

then we can see that the middle number of these is 32, i.e. the median is 32.

If there is an even number of values, the median is found by finding the average or mean of the two middle values after they have been placed in ascending or descending order.

To find the median of 24, 32, 36, 29, 31, 34, 35, 39, rearrange in ascending order:

$$24, 29, 31, 32, 34, 35, 36, 39$$

Then the median is $\dfrac{32 + 34}{2} = \dfrac{66}{2} = 33$ i.e. the median is 33.

EXERCISE 19E

Find the median of each of the following sets of numbers.

1 1, 2, 3, 5, 7, 11, 13

2 26, 33, 39, 42, 64, 87, 90

3 13, 24, 19, 13, 6, 36, 17

4 4, 18, 32, 16, 9, 7, 29

5 1.2, 3.4, 3.2, 6.5, 9.8, 0.4, 1.8

6 5, 7, 11, 13, 17, 19

7 34, 46, 88, 92, 104, 116, 118, 144

8 34, 42, 16, 85, 97, 24, 18, 38

9 1.92, 1.84, 1.89, 1.86, 1.96, 1.98, 1.73, 1.88

This table shows the marks of Class 7G in a maths test.

Mark	Frequency
0	1
1	5
2	7
3	8
4	8
5	4
	Total: 33

a Find the median mark.

b Lisa got 4 for this test. Is it true to say that she did better than half the class?

a There are 33 marks so the median mark is the 17th mark.

> We can count down the frequencies until we find the 17th mark. The first 6 marks are zero and ones, adding on 7 gives 13 marks and takes us to the end of the twos. Adding on 8 gives 21 marks and takes us to the end of the threes. Therefore the 17th mark is in the list of threes, so 3 is the median mark.

The median mark is 3.

b Yes because Lisa's mark is higher than the median mark.

These distributions come from Exercise 19C. Find the median of each set of values.

10 The pupils in Class 7G gathered this information about themselves.
Find the median number of children per family.

Number of children in each family	Frequency
1	8
2	12
3	4
4	2

11 Joshua tossed three coins several times and recorded the number of heads that showed at each toss. His results are shown in the table.
Find the median number of heads per toss.

Number of heads obtained when three coins are tossed	Frequency
0	9
1	7
2	16
3	3

12 Once every five minutes Debbie counted the number of people queuing at a checkout and gave her results in this table.
What is the median number of people queueing?

Number of people queuing at a supermarket checkout	Frequency
0	4
1	6
2	5
3	2
4	2

MODE

The mode of a set of numbers is the number that occurs most frequently, e.g. the mode of the numbers

$$6, 4, 6, 8, 10, 6, 3, 8 \text{ and } 4$$

is 6, since 6 is the only number occurring more than twice.

It would obviously be of use for a firm with a chain of shoe shops to know that the mode or modal size for men's shoes in one part of the country is 8, whereas in another part of the country it is 7. Such information would influence the number of pairs of shoes of each size kept in stock.

If all the figures in a set of figures are different, there cannot be a mode, for no figure occurs more frequently than all the others. On the other hand, if two figures are equally the most popular, there will be two modes.

In Chapter 3, we used bar charts to show such things as the spread of shoe sizes in a group of children, and the favourite colour of a group of people. These may be used to determine the mode of the group.

The following bar chart shows the colour selected by 35 people when asked to choose their favourite colour from a card showing six colours.

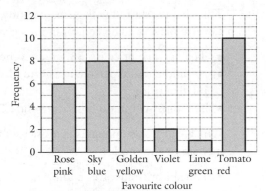

It shows that the most popular colour, or the modal colour, is tomato red.

EXERCISE 19F

What is the mode of each of the following sets of numbers?

1 10, 8, 12, 14, 12, 10, 12, 8, 10, 12, 4

2 3, 9, 7, 9, 5, 4, 8, 2, 4, 3, 5, 9

3 1.2, 1.8, 1.9, 1.2, 1.8, 1.7, 1.4, 1.3, 1.8

4 58, 56, 59, 62, 56, 63, 54, 53

5 5.9, 5.6, 5.8, 5.7, 5.9, 5.9, 5.8, 5.7

6 26.2, 26.8, 26.4, 26.7, 26.5, 26.4, 26.6, 26.5, 26.4

7 The table shows the number of goals scored by a football club last season.

Number of goals	0	1	2	3	4	5	6
Frequency	12	16	7	4	2	0	1

Draw a bar chart to show these results and give the modal score.

8 Given below are the marks out of 10 obtained by 30 girls in a history test.

8, 6, 5, 7, 8, 9, 10, 10, 3, 7, 3, 5, 4, 8, 7, 8, 10, 9, 8, 7, 10, 9, 9, 7, 5, 4, 8, 1, 9, 8

Draw a bar chart to show this information and find the mode.

9 The heights of 10 girls, correct to the nearest centimetre, are

155, 148, 153, 154, 155, 149, 162, 154, 156, 155

What is their modal height?

WHICH MEASURE
OF CENTRAL
TENDENCY ?

This chapter started with Aisha wanting to know about prices of personal cassette players. The central figures given are

- The average price is £35. This is the mean price.
- A middle range personal cassette player costs about £30. This is the median price.
- The most common cost is around £20. This is the modal price.

Which of these summaries is most useful to start with – the mean, the median or the mode ?

The mean price involves adding up all the prices so its value takes account of the very expensive machines, and she would probably not be interested in these.

The median price shows that she has a choice of half the models available for a cost of £30 or less, so this is quite useful.

The modal price shows that there is more than one model costing about £20 so there is a choice of models at this price. It doesn't tell us how many models are priced at about £20 – it could be only two or three, so this information is of limited use.

EXERCISE 19G

Discuss with the class the most useful way of summarising the information for the purpose given.

1 You have a list of the heights of 100 twelve-year-old girls. You want to know if you are taller or shorter than most of these girls.

2 You have a list giving the numbers sold over the last term of each item stocked by the school tuck shop.

   a You need to buy more stock and do not want to run out of the most popular item.

   b You have to buy the same quantity of each item.

3 You have a list giving the maximum and minimum temperature in Cairo each day in February for the last five years. You need to decide what clothes to take with you for a 10-day trip to Cairo in February.

4 You have a list of all models of inkjet printers that will print colour on A4 paper. You want an idea of the price you would have to pay for one.

5 Which form of summary would be most useful for each situation given in Exercise 19A ?
Give a reason for your choice.

1 In seven rounds of golf, a golfer returns scores of:

72, 87, 73, 72, 86, 72 and 77

Find the mean, mode and median of these scores.

2 The heights (correct to the nearest centimetre) of a group of boys are

159, 155, 153, 154, 157, 162, 152, 160, 161, 157

Find   **a** their mean height   **c** their median height
   b their modal height   **d** the range of the heights.

3 The table shows how many pupils in a form were absent for various numbers of sessions during a certain school week.

Number of sessions absent	0	1	2	3	4	5	6	7	8	9	10
Frequency	20	2	4	0	2	0	1	2	0	0	1

Find   **a** the mode   **b** the median   **c** the mean.

4 The mean number of words in each sentence on the first page of *The Machine-Gunners* is 10 and the lengths of sentences range from one word to 25 words. For the first page of *Oliver Twist*, the mean length of the sentences is 68 words and they range from 34 words to 98 words.
Use this information to compare the sentences in each book.

5 The table shows the marks obtained by a group of children in a quiz.

Mark	1	2	3	4	5	6	7	8	9	10
Frequency	5	10	14	10	9	8	8	6	6	9

a How many children took part in the quiz?

b Find the mean, median and modal marks.

Which of the following statements are true?

c More than half the children got a score of 4 or more.

d A score of 3 is the most likely score.

e Jane got a score less than the mean, so more than half the group did better than Jane.

6 These frequency tables show the number of 20 p coins that ten children from two different classes had with them on one day.

Class 7T	
Number of 20 p coins	Frequency
0	2
1	5
2	0
3	2
4	0
5	0
6	0

Class 7R	
Number of 20 p coins	Frequency
0	0
1	4
2	3
3	2
4	0
5	0
6	1

a Find the mean number of coins that each group had with them.

b Is the mean a good representative for each distribution?

c What would you use if you wanted to compare the two distributions?

PRACTICAL WORK

1 If you throw a fair dice 60 times, then *in theory* you should get each score 10 times, i.e. a frequency table would look like this.

Score	1	2	3	4	5	6
Frequency	10	10	10	10	10	10

a Find the mean score of this theoretical distribution.

b If you throw a dice 60 times, you will be very unlikely indeed to get equal numbers of each score. But, if your dice is unbiased, and if you throw it fairly, you should get a mean score near to the theoretical mean.

Throw a dice 60 times, record your results and find your mean score. How does your mean compare with the theoretical mean?

Can you say whether your dice is likely to be unbiased?

c Now repeat part **b** with an obviously biased dice. (You can make one like the one shown on page 376 and stick a bit of Blu-Tack inside one face before you make it up.)

2 This information is from an Argos catalogue. It gives details of prices of personal cassette players.

a What is the cheapest player shown here?

b What is the range of prices? **c** Is there a mode?

d Find the median price and the mean price.

e Compare the information given here with the information at the start of this chapter.

SOLIDS

We live in a three-dimensional world among solid objects. Solids have depth as well as length and breadth. Sometimes we have to tell other people what a particular solid looks like; we might, for example, see a bookcase in a shop and want to describe it to someone. We could use words, or a drawing or a photograph; we cannot show the bookcase itself unless we buy it or take people to the shop to look at it.

Sometimes there is no object to show, just an idea of one in our heads; Greg is designing a set of building blocks for young children.
To start with, he could

- make a few blocks
- draw his ideas on paper

Trying out ideas by making blocks is time consuming. It is better to try them out by drawing them on paper first.

EXERCISE 20A

1 Work in pairs for this question.

a One of you should think of a table in your home. Try describing it to your partner in words, without using the word 'table'.

b Now try drawing the table.
Which method gives most information?

c Now change roles and repeat parts **a** and **b**.

2 Lydia has an idea for a new desk which she is hoping to be allowed to make for herself.
Discuss thee ways in which she can get her ideas over to her family.

Drawing a three-dimensional object on paper is often the most effective way of showing other people what it looks like.

DRAWING CUBES AND CUBOIDS

A cube is a three-dimensional object. Each of its faces is a square. When we draw a cube on a flat sheet of paper only one of its faces is drawn as a square, the others that can be seen are drawn as parallelograms. Using a square grid of lines or dots makes it easier to get parallel edges.

Rectangular blocks, called cuboids, can be drawn in a similar way.

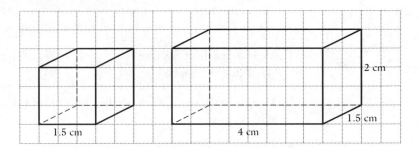

EXERCISE 20B

1 Use the diagrams above to answer the following questions.

  a Measure each of the lines on the drawing of the cube.
  Are they all the same length?

  b Measure each of the lines on the drawing of the cuboid.
  Are these the same as the measurements marked on the drawing?

2 Use squared paper to draw a cube of side 3 cm.
  Do you need to put the measurements on the diagram?
  Give a reason for your answer.

3 Use squared paper to draw a picture of a cuboid measuring 4 cm by
  2 cm by 1 cm.
  If you do not put the dimensions of the cuboid on your drawing, can
  someone else tell what the measurements of the cuboid are?

**USING
ISOMETRIC
PAPER TO DRAW
CUBES AND
CUBOIDS**

A cube can be drawn in another way, with a vertical edge right at the
front. Drawn in this way, each face is a parallelogram. This is not easy to
do freehand but there is a grid, called an isometric grid, that makes the
job simple. (Sometimes the grid is replaced by dots.)
Cuboids can be drawn in a similar way.

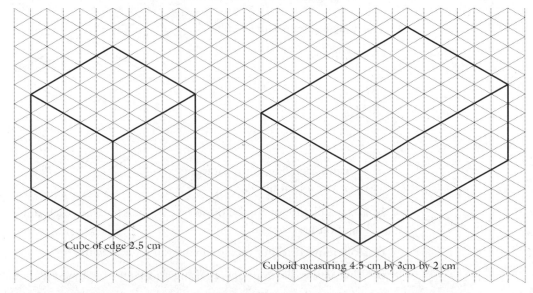

Cube of edge 2.5 cm

Cuboid measuring 4.5 cm by 3cm by 2 cm

EXERCISE 20C   Use isometric paper for this exercise. Make sure that the paper is the correct way round, i.e. that one set of lines (or dots) is vertical.

1 Measure each of the lines on the drawings of the cubes on page 373. What do you notice ?

2 Measure each of the lines on the drawings of the cuboids on page 373. What do you notice ?

3

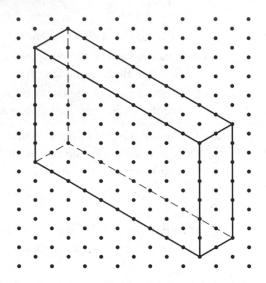

This drawing represents a cuboid measuring 5 cm by 3 cm by 1 cm.

a Measure each line on the drawing. What do you notice ?

b Draw a picture of a cuboid measuring 6 cm by 4 cm by 3 cm. Measure each of your lines. What do you notice ?

4 This drawing represents a cube of side 2 cm.

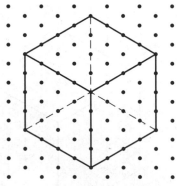

a Measure each line in this drawing. What do you notice ?

b Draw a picture of a cube of side 4 cm, then measure each of the lines you have drawn.

c Do you need to put measurements on your drawing ?

d What is the disadvantage of using this paper to draw a cube ?

When cubes and cuboids are drawn on isometric paper,
the lengths of the lines representing the edges are
the correct length.

5 Draw a picture of each of the following solids.

a A cube of side 5 cm   **b** A cube of side 8 cm

c A cuboid 6 cm by 4 cm by 1 cm

d A cuboid 3 cm by 5 cm by 3 cm

6 Isometric paper is useful when
drawing stacks of cubes.

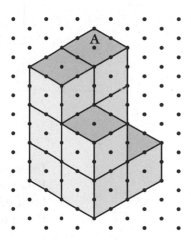

a How many loose cubes are
needed to make the stack that
you can see here?

b Explain how your answer to
part **a** might be affected if you
are told that the cubes are stuck
together.

c Draw the stack with the cube
marked **A** removed.

**MAKING CUBES
AND CUBOIDS**

Any solid with flat faces can be made from a flat sheet.
(We are using the word 'solid' for any object that takes up space, i.e. for
  any three-dimensional object, and such an object can be hollow.)

A cube can be made from six separate squares.

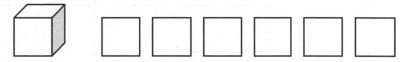

We can avoid a lot of unnecessary sticking if we join some squares
together before cutting out.

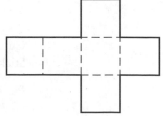

This is called a *net*.

There are other arrangements of six squares that can be folded up to
make a cube. Not all arrangements of six squares will work however, as
we will see in the next exercise.

1 Below is the net of a cube of edge 4 cm.

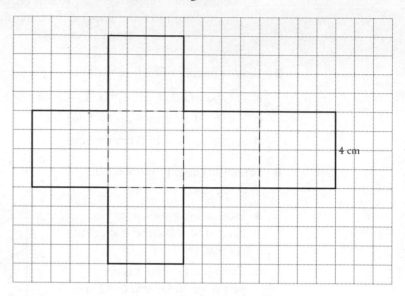

Draw the net on 1 cm squared paper and cut it out.
Fold it along the broken lines.
Fix it together with sticky tape.
If you mark the faces with the numbers 1 to 6, you can make a dice.

2 Draw this net full-size on 1 cm squared paper.

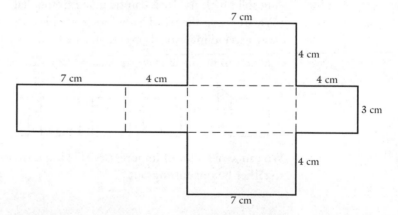

Cut the net out and fold along the dotted lines.
Stick the edges together.

a **i** How many faces are rectangles measuring 7 cm by 4 cm?
  ii How many faces are rectangles measuring 7 cm by 3 cm?
  iii What are the measurements of the remaining faces?

b Draw another arrangement of the rectangles which will fold up to make this cuboid.

3 This cuboid is 4 cm long, 2 cm wide and 1 cm high.

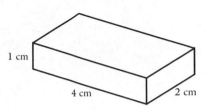

 a How many faces does this cuboid have?

 b Sketch the faces, showing their measurements.

 c On 1 cm grid paper, draw a net that will make this cuboid.

4 This net will make a cuboid.

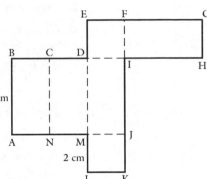

 a Sketch the cuboid, and show its measurements.

 b Which edge meets HI?

 c Which other corners meet at A?

5 This net will make a cube.

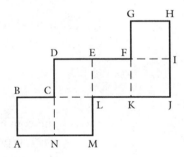

 a Which edge meets AB?

 b Which other corners meet at H?

6

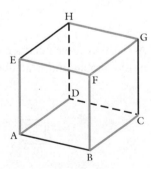

This cube is cut along the edges drawn with a coloured line and flattened out.

Draw the flattened shape.

7 Here are two arrangements of six squares.

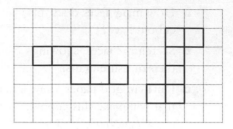

    a Copy these on 1 cm squared paper.

    b Draw as many other arrangements of six squares as you can find.

    c Which of your arrangements, including the two given here, will fold up to make a cube?
    If you cannot tell by looking, cut them out and try to make a cube.

VOLUME

The volume of a solid is the amount of space it occupies.

In a science laboratory you may have seen a container with a spout, similar to the one shown in the diagram.

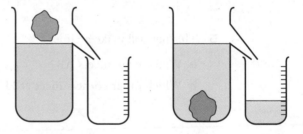

The container is filled with water to the level of the spout. Any solid which is put into the water will force a quantity of water into the measuring cylinder. The volume of this water is equal to the volume of the solid.

CUBIC UNITS

As with area, we need a convenient unit for measuring volume. The most suitable unit is a cube.

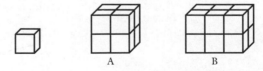

Now we can count how many of the smallest cubes are needed to fill the same space as each of the solids A and B: 8 small cubes fill the same space as solid A and 12 small cubes fill the same space as solid B.

This tells us that solid B occupies $1\frac{1}{2}$ times as much space as solid A, but it does not help if we want to compare the volume of A with the volume of this book, say. To do this we need to measure volume in standard cubic units. These are based on the standard units of length.

A cube with a side of 1 cm has a volume of one cubic centimetre which is written $1\,\text{cm}^3$.

Similarly a cube with a side of 1 mm has a volume of $1\,\text{mm}^3$ and a cube with a side of 1 m has a volume of $1\,\text{m}^3$.

VOLUME OF A CUBOID

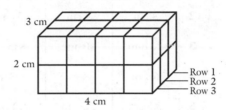

The diagram shows a cuboid measuring 4 cm by 3 cm by 2 cm.
To cover the area on which the block stands we need three rows of standard 1 cm cubes with four cubes in each row, i.e. 12 cubes.
A second layer of 12 cubes is needed to give the volume shown, so the volume of the block is 24 standard 1 cm cubes.

Therefore the volume of the solid is $24\,\text{cm}^3$.

This is also given when we calculate length × breadth × height,
i.e.   the volume of the block $= 4 \times 3 \times 2\,\text{cm}^3$

or   the volume of the cuboid $=$ length × breadth × height

EXERCISE 20E

1 Which unit would you use to give the volume of

   a this book

   b the room you are in

   c one vitamin pill

   d a lorry load of rubble

   e a packet of cornflakes

   f a 2 p coin

   g a concrete building block?

Find the volume of a cuboid measuring 12 cm by 10 cm by 5 cm.

$$\text{Volume of cuboid} = \text{length} \times \text{breadth} \times \text{height}$$
$$= 12 \times 10 \times 5 \ \text{cm}^3$$
i.e. $\qquad \text{Volume} = 600 \ \text{cm}^3$

Find the volume of each of the following cuboids.

	Length	Breadth	Height
2	4 cm	4 cm	3 cm
3	20 mm	10 mm	8 mm
4	6.1 m	4 m	1.3 m
5	3.5 cm	2.5 cm	1.2 cm
6	4 m	3 m	2 m
7	8 cm	5 m	4 m
8	8 m	3 cm	$\frac{1}{2}$ cm
9	12 cm	1.2 cm	0.5 cm
10	4.5 m	1.2 m	0.8 m

When the lengths of the edges of a cuboid are given in different units, we must change some of them so that all measurements are given in the same unit.

11 A rectangular block of concrete is 1 m long, 20 cm wide and 25 cm deep.
Find the volume of the block in cubic centimetres.

12 A rectangular piece of aluminium measures 25 cm by 15 mm by 5 mm.
Find its volume in cubic millimetres.

13 The shape of an outdoor aviary is a cuboid with measurements 2 m by 90 cm by 90 cm.
Find the volume of the aviary giving your answer in the most appropriate unit.

Find the volume of a cube with edge 6 cm.

$$\text{Volume of cube} = \text{length} \times \text{breadth} \times \text{height}$$
$$= 6 \times 6 \times 6 \text{ cm}^3$$

i.e. $\quad\quad\quad$ Volume $= 216 \text{ cm}^3$

Find the volume of a cube with the given side.

14 4 cm $\quad\quad\quad\quad$ **17** $\frac{1}{2}$ cm $\quad\quad\quad\quad$ **20** 8 cm

15 5 cm $\quad\quad\quad\quad$ **18** 2.5 cm $\quad\quad\quad\quad$ **21** $\frac{1}{2}$ m

16 2 m $\quad\quad\quad\quad$ **19** 3 km $\quad\quad\quad\quad$ **22** 3.4 m

Draw a cube of side 8 cm.
How many cubes of side 2 cm would be needed to fill the same space?

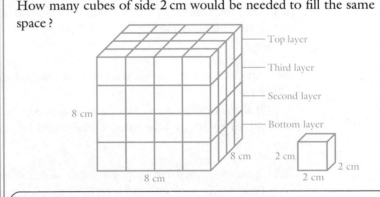

The bottom layer needs 4 × 4, i.e. 16 cubes of side 2 cm, and there are four layers altogether.

64 cubes are required.

23 Draw a cube of side 4 cm.
How many cubes of side 2 cm would be needed to fill the same space?

24 Draw a cuboid measuring 6 cm by 4 cm by 2 cm.
How many cubes of side 2 cm would be needed to fill the same space?

25 Draw a cube of side 6 cm.
How many cubes of side 3 cm would be needed to fill the same space?

26 Draw a cuboid measuring 8 cm by 6 cm by 2 cm.
How many cubes of side 2 cm would be needed to fill the same space?

27 A cuboid measures 4 cm by 8 cm by 4 cm. The same space is to be filled with smaller cubes.
How many cubes are needed if their sides are

   a .1 cm   **b** 2 cm   **c** 4 cm?

28 Find the volume of air in a room measuring 4 m by 5 m which is 3 m high.

29 Find the volume, in cm^3, of a concrete block measuring 36 cm by 18 cm by 12 cm.

30 Find the volume of a school hall which is 30 m long and 24 m wide if the ceiling is 9 m high.

31 An electric light bulb is sold in a box measuring 10 cm by 6 cm by 6 cm.
If the shopkeeper receives them in a carton measuring 50 cm by 30 cm by 30 cm, how many bulbs would be packed in a carton?

32 A classroom is 10 m long, 8 m wide and 3 m high.
How many pupils should it be used for if each pupil requires 5 m^3 of air space?

33 How many rectangular packets, measuring 8 cm by 6 cm by 4 cm, can be packed in a rectangular cardboard box measuring 30 cm by 24 cm by 16 cm?

34 A kitchen worktop is 2 metres long, 60 cm deep and 90 cm above the floor.
What is the volume of the space below the worktop?

35 Rectangular blocks of stone measure 30 cm by 20 cm by 35 mm.
How many of these blocks can be laid flat in a cube-shaped crate whose edge is 1 metre?

CHANGING UNITS OF VOLUME

Consider a cube of side 1 cm. If each edge is divided into 10 mm the cube can be divided into 10 layers, each layer with 10×10 cubes of side 1 mm,

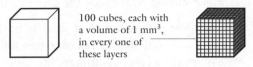

100 cubes, each with a volume of 1 mm³, in every one of these layers

i.e.

$$1 \text{ cm}^3 = 10 \times 10 \times 10 \text{ mm}^3$$
$$= 1000 \text{ mm}^3$$

Similarly, since $1 \text{ m} = 100 \text{ cm}$

i.e.

$$1 \text{ cubic metre} = 100 \times 100 \times 100 \text{ cm}^3$$
$$= 1\,000\,000 \text{ cm}^3$$

EXERCISE 20F

Express   **a** 2.4 m^3 in cm^3 **b** 0.0025 cm^3 in mm^3
         c 126 cm^3 in m^3

a Since $1 \text{ m}^3 = 100 \times 100 \times 100 \text{ cm}^3$
       $2.4 \text{ m}^3 = 2.4 \times 100 \times 100 \times 100 \text{ cm}^3$
            $= 2\,400\,000 \text{ cm}^3$

b As $1 \text{ cm}^3 = 10 \times 10 \times 10 \text{ mm}^3$
    $0.0025 \text{ cm}^3 = 0.0025 \times 10 \times 10 \times 10 \text{ mm}^3$
             $= 2.5 \text{ mm}^3$

c $1 \text{ m}^3 = 1\,000\,000 \text{ cm}^3$ so $1 \text{ cm}^3 = \frac{1}{1\,000\,000} \text{ m}^3$
   $\therefore$ $126 \text{ cm}^3 = 126 \div 1\,000\,000 \text{ m}^3$
             $= 0.000\,126 \text{ m}^3$

Express in mm^3

1 8 cm^3 **3** 6.2 cm^3 **5** 0.0092 m^3

2 14 cm^3 **4** 0.43 cm^3 **6** 0.0004 cm^3

Express in cm^3

7 3 m^3 **10** 0.0063 m^3 **13** $29\,300 \text{ mm}^3$

8 2.5 m^3 **11** 22 mm^3 **14** 2.5 mm^3

9 0.42 m^3 **12** 731 mm^3 **15** 1000 mm^3

CAPACITY

When we buy a bottle of milk or a can of engine oil we are not usually interested in the external measurements or volume of the container. What really concerns us is how much milk is inside the bottle, or how much engine oil is inside the can, i.e., the *capacity* of the container.

The most common unit of capacity in the metric system is the litre. (A litre is roughly equivalent to two full bottles of milk.) A litre is much larger than a cubic centimetre but much smaller than a cubic metre. The relationship between these quantities is:

$$1000 \text{ cm}^3 = 1 \text{ litre}$$

i.e. a litre is the volume of a cube of side 10 cm.

$$1000 \text{ litres} = 1 \text{ m}^3$$

When the amount of liquid is small, such as dosages of medicines, the millilitre (ml) is used. A millilitre is a thousandth part of a litre, i.e.

$$1000 \text{ ml} = 1 \text{ litre} \quad \text{or} \quad 1 \text{ ml} = 1 \text{ cm}^3$$

EXERCISE 20G

Express 5.6 litres in cm^3

$$1 \text{ litre} = 1000 \text{ cm}^3$$
$$5.6 \text{ litres} = 5.6 \times 1000 \text{ cm}^3$$
$$= 5600 \text{ cm}^3$$

Express in cm^3

1 2.5 litres   **3** 0.54 litres   **5** 35 litres

2 1.76 litres   **4** 0.0075 litres   **6** 0.028 litres

Express in litres

7 7000 cm^3 **8** 4000 cm^3 **9** 2400 cm^3

Express in litres

10 5 m^3 **11** 12 m^3 **12** 4.6 cm^3

Find the capacity, in litres, of a water tank whose internal measurements are 2 m by 70 cm by 30 cm.

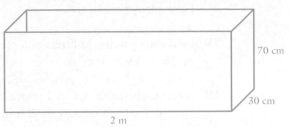

We need to find the volume of the tank in cm³. To do this, all the measurements must be in centimetres so first convert the 2 m into centimetres.

Length of cuboid $= 2\,\text{m} = 2 \times 100\,\text{cm} = 200\,\text{cm}$
Volume of cuboid $=$ length $\times$ breadth $\times$ height
$$= 200 \times 70 \times 30\,\text{cm}^3$$
$$= 420\,000\,\text{cm}^3$$

$1000\,\text{cm}^3 = 1$ litre, so to convert cubic centimetres to litres, we divide by 1000.

Capacity of tank $= 420\,000 \div 1000$ litres
$$= 420 \text{ litres.}$$

13 Find the capacity, in litres, of these cuboids.

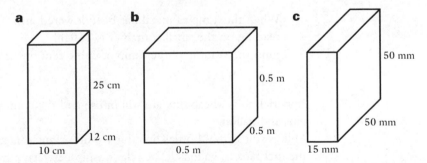

a   25 cm   12 cm   10 cm

b   0.5 m   0.5 m   0.5 m

c   50 mm   50 mm   15 mm

14 How many litres will it take to fill a rectangular petrol can measuring 30 cm by 20 cm by 10 cm?

15 A metal block, measuring 50 cm by 12 cm by 12 cm is melted. How many litres of liquid metal are there?

16 A rectangular water storage tank is 3 m long, 2 m wide and 1 m deep. How many litres of water will it hold?

17 How many cubic metres of water are required to fill a rectangular swimming bath 15 m long and 10 m wide which is 2 m deep throughout?
How many litres is this?

18 Find the capacity, in litres, of a rectangular carton measuring 20 cm by 15 cm by 30 mm.

19 A rectangular fish tank is 1 metre long, 30 cm deep and 20 cm high. How many litres of water will it hold?

20 A rectangular carton of concentrated orange juice measures 5 cm by 10 cm by 4 cm. To make one glass of juice, 5 ml of this concentrate are needed. How many glasses of juice can be made from one full carton?

21 How many lead cubes, of side 2 cm, can be cast from 5 litres of liquid lead?

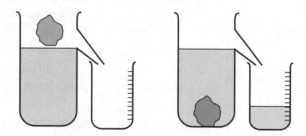

When the lump of metal has been lowered into the water, the reading on the small container is 250 ml.
Find the volume of the lump in cubic centimetres.

IMPERIAL UNITS OF CAPACITY

Imperial units of capacity are still in use and the common ones are the pint and gallon.
Milk is still sold in bottles holding 1 pint. Many cars give the capacity of the fuel tank in gallons. The relationship between pints and gallons is

1 gallon = 8 pints

Approximate conversions between metric and imperial units of capacity can be made using

1 litre ≈ 1.75 pints   and   1 gallon ≈ 4.5 litres

EXERCISE 20H

1 Give, roughly, the number of pints equivalent to

    a 20 litres   **b** 12 litres   **c** 1.5 litres

2 Roughly, how many gallons is

    a 50 litres   **b** 30 litres   **c** 25 litres?

3 Give the approximate number of litres equivalent to

    a 4 pints   **b** $2\frac{1}{2}$ gallons   **c** 10 gallons

4 Arrange these containers in order of capacity, with the largest first.

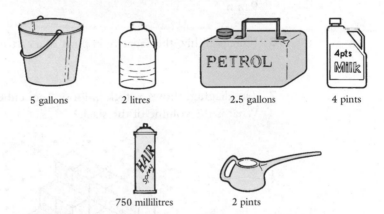

5 gallons   2 litres   2.5 gallons   4 pints

750 millilitres   2 pints

5 Liz fills her car up with petrol. The reading on the pump is 38 litres. How many gallons is that?

6 A recipe needs 100 ml of cream. Will a half-pint carton be enough?

7 A rectangular fish tank measures 300 mm by 250 mm by 500 mm. Roughly, how many gallons of water will it hold when full?

8 Petrol costs 56 p a litre. How much, roughly, will it cost to fill a tank whose capacity is 1.5 gallons?

EXERCISE 20I

1 Express $3.2 \, \text{m}^3$ in   **a** cm^3 **b** mm^3

2 Express 1.6 litres in cm^3.

3 Find the volume of a cube of side 4 cm.

4 Find the volume, in cm^3, of a cuboid measuring 2 m by 25 cm by 10 cm.

5 Find the volume, in mm^3, of a cuboid measuring 5 cm by 3 cm by 9 mm.

6 Find, roughly, the capacity in pints of a carton designed to hold 1.5 litres.

7 The diagram shows a stack of loose 2 cm cubes. What is the volume of the stack?

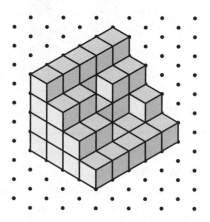

EXERCISE 20J

1 Express $8 \, \text{cm}^3$ in   **a** mm^3 **b** m^3

2 Express $3500 \, \text{cm}^3$ in litres.

3 Find the volume of a cuboid measuring 10 cm by 5 cm by 6 cm.

4 Find, in cm^3, the volume of a cube of side 8 mm.

5 Find the volume, in cm^3, of a cuboid measuring 50 cm by 1.2 m by 20 cm.

6 Which holds more water, a rectangular carton measuring 20 cm by 20 cm by 10 cm or a bottle whose capacity is 750 ml?

7 Which holds more liquid, a plastic carton whose capacity is 1 pint or a plastic bottle whose capacity is 500 ml?

8 Find the volume of this stack of 1 cm cubes.

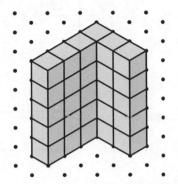

PUZZLE

Which two of these shapes will fit together to make a cube?

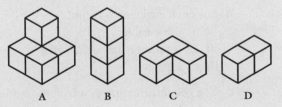

A   B   C   D

PRACTICAL WORK

This pattern is drawn on isometric dots. It can be continued in the same way to fill the grid.

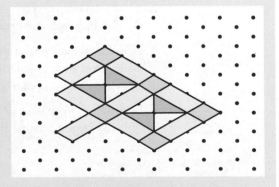

Draw some different patterns on isometric paper.

INVESTIGATION

A rectangular piece of card measuring 12 cm by 9 cm is to be used to make a small open rectangular box. The diagram shows one way of doing this.

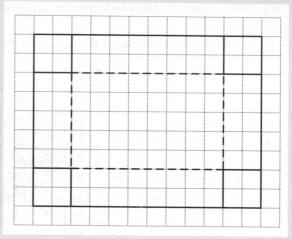

a How many different open boxes can be made from this piece of card if the length and breadth of the base is to be a whole number of centimetres?

Draw each possibility on squared paper.

b For each different box find
   i its dimensions
   ii its total external surface area
   iii its capacity.

c Use the information you have found in part **b** to state whether each of these statements is true (T) or false (F).
   A The box with the largest base has the greatest capacity.
   B The box with the smallest total external surface area has the smallest capacity.
   C The box with the greatest capacity is as deep as it is wide.

d Find the dimensions of the largest cubical open box that can be made from a rectangular card measuring 12 cm by 9 cm.
   How much card is wasted?

e Investigate other sizes of card to find the dimensions of the box with the largest volume that can be made from each size.

EQUATIONS

Jo went into a shop and bought some chocolate for Karl and some boiled sweets for herself. Together they cost £1.13. When Karl asked Jo how much he owed her for his chocolate all she could remember was that it cost 15 p more than her boiled sweets. They could find out how much each had to pay by

- guessing until they found two amounts that have a total of 113 p and a difference of 15 p, or
- writing down the information using mathematical symbols.

To do this we can write

$$(\text{amount Karl owed}) + (\text{amount Jo owed}) = 113 \, \text{p}$$

but (amount Jo owed) = (amount Karl owed − 15 p)

so (amount Karl owed) + (amount Karl owed − 15 p) = 113 p

This is called an *equation*.
Now we can see that twice the amount Karl owed = 128 p
so Karl owed Jo 64 p.

EXERCISE 21A

Discuss how you might solve the problems that arise in the following situations.

1 Chris and Joy have been given £10 between them and must divide it so that Chris has £1 more than Joy.

2 Rod and Daniel go for a meal. The total cost comes to £34.60. Daniel's meal costs £1.70 more than Rod's and Rod is unwilling to split the bill down the middle. He wants each of them to pay the cost of his own meal.

3 A teacher had a full box of 100 drawing pins. She gave 4 to each pupil to pin a painting on the wall. She came across 3 faulty pins and still had 5 pins left over when all the pupils had taken the pins they needed. She wondered how many pupils had come to her to put their paintings up.

4 Can you think of other situations where it should be easier to form an equation and use it, rather than guess and then have to check to see if the chosen solution works?

I think of a number, add seven and get twenty-one. This sentence could be rewritten

$$\text{The number I think of} + 7 = 21$$

from which we can see that the number I think of is 14.

If we use a letter (we shall use x) to stand for the number I think of, the sentence could be rewritten

$$x + 7 = 21$$

Then, if we take away 7 from each side

$$x = 21 - 7$$
$$= 14$$

so the number I first thought of was 14.

> I think of a number, add 4, and the result is 10.
> If x stands for the unknown number, form an equation in x and solve it to find the number I thought of.
>
> The equation is $x + 4 = 10$
> The number is 6

In each question from **1** to **15** let the letter x stand for the unknown number.

Use the given statement to form an equation in x and solve the equation to find the unknown number.

1 I think of a number, subtract 3 and get 4.

2 I think of a number, add 1 and the result is 3.

3 If a number is added to 3 we get 9.

4 If 5 is subtracted from a number we get 2.

5 I think of a number, add 8 and get 21.

6 If 7 is subtracted from a number we get 19.

I think of a number, multiply it by 3 and the result is 12.
What is the number?

The equation is $3x = 12$
The number is 4.

$3x$ means $3 \times x$

7 I think of a number, double it and get 8.

8 If a number is multiplied by 7 the result is 14.

9 When we multiply a number by 3 we get 15.

10 6 times an unknown number gives 24.

11 I think of a number, multiply it by 4 and get 24.

12 I think of a number, divide it by 6 and get 5.

13 When a number is divided by 3 we get 7.

14 If a number is divided by 7 the answer is 7.

15 I think of a number, divide it by 8 and get 32.

Write a sentence to show the meaning of the equation $4x = 20$.

$4x = 20$ means '4 times an unknown number gives 20', or,
'I think of a number, multiply it by 4 and the result is 20'.

Write sentences to show the meaning of the following equations.

16 $3x = 18$ **20** $5 + x = 7$

17 $x + 6 = 7$ **21** $x - 4 = 1$

18 $x - 2 = 9$ **22** $4x = 8$

19 $5x = 20$ **23** $x + 1 = 4$

**SOLVING
EQUATIONS**

Some equations need an organised approach, rather than guesswork.

Imagine a balance:

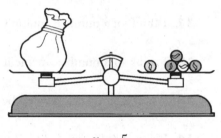

On this side there
is a bag containing
an unknown number
of marbles,
say x marbles,
and 4 loose marbles.

On this side
there are 9 separate
marbles, balancing
the marbles on the
other side.

$$x + 4 = 9$$

Take 4 loose marbles from each side, so that the two sides still balance.

$$x = 5$$

We write: $x + 4 = 9$

Take 4 from both sides $x = 5$

When we have found the value of x we have *solved the equation.*

As a second example suppose that :
$$x - 2 = 5$$

On this side there is
a bag that originally
held x marbles but
now 2 are missing.

On this side there
are 5 loose
marbles.

We can make the bag complete by putting back 2 marbles but, to keep the balance, we must add 2 marbles to the right-hand side also.

So we write $\qquad x - 2 = 5$

Add 2 to each side $\qquad x = 7$

> Whatever you do to one side of an equation you must also do to the other side.

Use the equations in the following exercise to practise this idea. It is easy enough to 'see' the solutions, so you should have no difficulty in checking your answers.

EXERCISE 21C

> Solve the equation $y + 4 = 6$
>
> $$y + 4 = 6$$
> Take 4 from both sides $\qquad y = 2$

Solve

1 $x + 7 = 15$ **5** $a + 3 = 7$ <u>**9**</u> $a + 1 = 6$

2 $x + 9 = 18$ **6** $x + 4 = 9$ <u>**10**</u> $a + 8 = 15$

3 $10 + y = 12$ **7** $a + 5 = 11$ <u>**11**</u> $7 + c = 10$

4 $2 + c = 9$ **8** $9 + a = 15$ <u>**12**</u> $c + 2 = 3$

13 Kim is n years old and Toni is 2 years older than Kim. If Toni is 14 we can form the equation

$$n + 2 = 14$$

How old is Kim?

14 A cup of coffee costs 15 pence more than a cup of tea. If a cup of tea costs x pence and a cup of coffee costs 75 pence we can form the equation

$$x + 15 = 75$$

How much is a cup of tea?

15 Adam has c cassettes and Peg has 8. Altogether they have 21 cassettes between them.
Use this information to form an equation.
How many cassettes does Adam have?

16 A concert hall has two car-parks and 364 cars are parked in them. There are 193 cars in the east car-park and n cars in the north car-park.

Use this information to form an equation in n.

How many cars are parked in the north car-park?

Some equations may have negative answers.

For example, if we know the temperature fell last night by $3\,°C$ to $-9\,°C$, we could find the temperature before it got colder by forming the equation

$$x - 3 = -9$$

where $x\,°C$ was the temperature before it fell.

Then adding 3 to both sides gives

$$x = -9 + 3 = -6$$

Solve $x + 8 = 6$

$$x + 8 = 6$$
Take 8 from both sides $x = -2$

Solve

17 $x + 4 = 2$	**19** $3 + a = 2$	**21** $4 + w = 2$
18 $x + 6 = 1$	**20** $s + 3 = 2$	**22** $c + 6 = 2$

Solve $x - 6 = 2$

$$x - 6 = 2$$
Add 6 to both sides $x = 8$

Solve

23 $x - 6 = 4$	**27** $c - 8 = 1$	**31** $a - 4 = 8$
24 $a - 2 = 1$	**28** $x - 5 = 7$	**32** $x - 3 = 0$
25 $y - 3 = 5$	**29** $s - 4 = 1$	**33** $c - 1 = 1$
26 $x - 4 = 6$	**30** $x - 9 = 3$	**34** $y - 7 = 2$

35 A chelsea bun costs 16 pence less than a jam doughnut. If a jam doughnut costs n pence and a chelsea bun 32 pence we can form the equation

$$n - 16 = 32.$$

What is the cost of a jam doughnut?

36 A small loaf weighs 450 grams less than a large loaf. If a large loaf weighs x grams and a small loaf weighs 500 grams, we can form the equation

$$x - 450 = 500$$

How heavy is a large loaf?

37 A cup of tea cost 80 p less than a sandwich. Together they cost 220 p.
Use this information to form an equation.
How much is a sandwich?

38 The temperature early this morning was $t\,°C$. Since then it has increased by 3° and is now $2\,°C$. From this information we can form the equation

$$t + 3 = 2$$

Solve this equation and so find the temperature early this morning.

39 Jean Pearce paid her gas bill for £80 by writing a cheque. Before it was presented at the bank she had a credit of £55. After it had been cleared she was £x overdrawn.
Form an equation in x and solve it.
How much overdrawn was she after the gas bill had been paid?

40 Simon Weinmann lives just outside Jerusalem. His house is 510 m above sea level. He also has a weekend flat near the Dead Sea which is 220 m below sea level. The vertical distance through which he descends when he travels from his Jerusalem home to his flat is x m.
Form an equation in x and solve it.
What vertical distance does he descend when he goes to his flat for the weekend?

EXERCISE 21D   Sometimes the letter term is on the right-hand side instead of the left.

Solve $3 = x - 4$

$$3 = x - 4$$

Add 4 to both sides $7 = x$ $7 = x$ is the same
$$x = 7$$ as $x = 7$

Solve

1 $4 = x + 2$ **3** $7 = a + 4$ **5** $1 = c - 2$

2 $6 = x - 3$ **4** $6 = x - 7$ **6** $5 = s + 2$

7 $x + 3 = 10$ **11** $6 + c = 10$ **15** $x - 6 = 5$

8 $9 + x = 4$ **12** $d + 4 = 1$ **16** $x + 3 = 15$

9 $c + 4 = 4$ **13** $7 = x + 3$ **17** $y - 6 = 4$

10 $3 = b + 2$ **14** $x + 1 = 9$ **18** $x - 7 = 4$

Solve $7.5 = x - 2.8$

$$7.5 = x - 2.8$$

Add 2.8 to both sides $7.5 + 2.8 = x$
i.e. $10.3 = x$
$$x = 10.3$$

Solve the equations.

19 $x - 1.5 = 6$ **22** $x - 3.2 = 5.6$ **25** $x - 4.1 = 7.8$

20 $6 = x - 4$ **23** $10 = a - 1$ **26** $x - 3 = 6$

21 $x - \frac{1}{2} = 5$ **24** $x + \frac{1}{2} = 4$ **27** $\frac{2}{3} = x - 1\frac{2}{3}$

28 $x - 4 = 2$ **31** $9 = x - 7$ **34** $c - 7 = 10$

29 $3 = x - 5.6$ **32** $x + 4.7 = 11.2$ **35** $6.6 = x - 3.9$

30 $4 + x = 5\frac{1}{4}$ **33** $\frac{3}{4} + x = 1\frac{1}{2}$ **36** $x - 1\frac{3}{4} = \frac{1}{2}$

37 $y - 9 = 14$ **40** $x + 1 = 8$ **43** $x + 8 = 1$

38 $2 = z - 2$ **41** $x - 1 = 8$ **44** $x - 8 = 1$

39 $d - 3 = 1$ **42** $1 = c + 3$ **45** $z + 3 = 5$

46 Terry cuts a piece of wood $1.75\,$m long from a piece that is $2.64\,$m long.
The piece he has left is $x\,$m long.
Form an equation in x and solve it.

47 A lorry weighs $5\frac{1}{4}$ tons and is loaded with $5\frac{3}{4}$ tons of steel. The total weight of the loaded lorry is N tons.
Form an equation in N and solve it to find the weight of the loaded lorry.

MULTIPLES OF x

Imagine that on this side of the scales there are 3 bags each containing an equal unknown number of marbles, say x in each.

On this side there are 12 loose marbles.

$$3 \times x = 12$$

$$3x = 12$$

We can keep the balance if we divide the contents of each scale pan by 3.

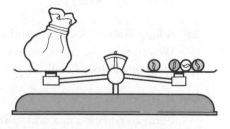

$$x = 4$$

EXERCISE 21E

Solve $6x = 12$

$$6x = 12$$
Divide both sides by 6 $\qquad x = 2$

Solve $3x = 7$

$$3x = 7$$
Divide both sides by 3 $\qquad x = \frac{7}{3}$

$$x = 2\tfrac{1}{3}$$

Solve the following equations.

1 $5x = 10$ $\qquad$ **5** $4b = 16$ $\qquad$ **9** $5p = 7$

2 $3x = 9$ $\qquad$ **6** $4c = 9$ $\qquad$ **10** $2x = 40$

3 $2x = 5$ $\qquad$ **7** $3a = 1$ $\qquad$ **11** $7y = 14$

4 $7x = 21$ $\qquad$ **8** $6z = 18$ $\qquad$ **12** $6a = 3$

13 $6x = 36$ $\qquad$ **17** $5z = 9$ $\qquad$ **21** $4y = 3$

14 $6x = 6$ $\qquad$ **18** $2y = 7$ $\qquad$ **22** $5x = 6$

15 $6x = 1$ $\qquad$ **19** $3x = 27$ $\qquad$ **23** $2z = 10$

16 $5z = 10$ $\qquad$ **20** $8x = 16$ $\qquad$ **24** $7x = 1$

25 Danny is n years old now. His father is 48 is four times as old as Danny. From this information we get the equation $4 \times n = 48$ or more simply $4n = 48$.
How old is Danny?

26 A light bulb costs x pence and a pack of 6 light bulbs costs 360 pence. We can form the equation $6x = 360$.
How much does 1 bulb cost?

27 A plank of wood 3.4 m long is cut into 4 equal pieces, each x m long. Form an equation in x and solve it.
How long is each piece?

28 Salman has $3\frac{1}{2}$ lb of potatoes. They are divided equally among 7 diners.

If each diner receives x lb form an equation in x and solve it.

What weight of potatoes does each diner receive?

MIXED OPERATIONS

EXERCISE 21F

Solve the following equations.

1 $x + 4 = 8$ **4** $5y = 6$ **7** $2x = 11$

2 $x - 4 = 8$ **5** $4x = 12$ **8** $x - 2 = 11$

3 $4x = 8$ **6** $x - 4 = 12$ **9** $x - 12 = 4$

10 $8 = c + 2$ **13** $7y = 2$ **16** $3 = a - 4$

11 $20 = 4x$ **14** $3x = 8$ **17** $3x = 5$

12 $2 + x = 4$ **15** $x + 6 = 1$ **18** $z - 5 = 6$

19 The temperature at midday was T °C. By 6 p.m. it had fallen by 6 °C to 12.5 °C.

From this information we can form the equation $T - 6 = 12.5$.

Solve this equation.

What was the temperature at midday?

Now that you think you know what the temperature was, read through the given information again to see that it fits.

20 A glass of squash costs 65 p and a chocolate eclair costs x pence more than a glass of squash. Together they cost 150 pence. As a chocolate eclair costs $(65 + x)$ pence we can form the equation $65 + 65 + x = 150$.

Solve this equation.

How much does an eclair cost?

21 Helen is x years old now. Her mother, who is 35 years old, is five times as old as Helen.

Form an equation in x and solve it.

How old is Helen now?

TWO OPERATIONS

One of the problems in the discussion exercise on page 391 concerned Rod and Daniel sharing the cost of a meal. If we suppose that Rod's meal cost £x then the cost of Daniel's meal was £$x + £1.70$. The total cost of the two meals was £34.60, so we can form the equation

$$x + x + 1.70 = 34.60$$

i.e. $2x + 1.70 = 34.60$

To solve this equation requires two operations.

First, subtract 1.70 from both sides

$$2x = 34.60 - 1.70$$

$$2x = 32.90$$

Secondly, divide both sides by 2

$$x = 16.45$$

The cost of Rod's meal was £16.45
and the cost of Daniel's meal was £16.45 + £1.70 = £18.15.
(*Check*:   cost of Rod's meal + cost of Daniel's meal
       = £16.45 + £18.15 = £34.60,
           which agrees with the information given in the question.)

EXERCISE 21G

Solve the equations   **a** $7 = 3x - 5$ **b** $2x + 3 = 5$

a $7 = 3x - 5$
Add 5 to both sides $12 = 3x$
Divide both sides by 3 $4 = x$ | To solve an equation our aim is to get the letter on its own. |
i.e. $x = 4$

b $2x + 3 = 5$
Take 3 from both sides $2x = 2$
Divide both sides by 2 $x = 1$

It is possible to check whether your answer is correct. We can put $x = 1$ in the left-hand side of the equation and see if we get the same value as on the right-hand side.

Check: If $x = 1$, left-hand side = $2 \times 1 + 3 = 5$
Right-hand side = 5, so $x = 1$ fits the equation.

Solve the following equations.

1 $6x + 2 = 26$ **7** $6 = 2x - 4$ **13** $20 = 12x - 4$

2 $4x + 7 = 19$ **8** $5z + 9 = 4$ **14** $9x + 1 = 28$

3 $17 = 7x + 3$ **9** $3x - 4 = 4$ **15** $9 = 8x - 15$

4 $4x - 5 = 19$ **10** $3x + 4 = 25$ **16** $8 = 8 + 3z$

5 $3a + 12 = 12$ **11** $13 = 3x + 4$ **17** $5x - 4 = 5$

6 $10 = 10x - 50$ **12** $5z - 9 = 16$ **18** $15 = 1 + 7x$

19 $9x - 4 = 14$ **25** $3a + 4 = 1$ **31** $10x - 6 = 24$

20 $3x - 2 = 3$ **26** $2x + 6 = 6$ **32** $5x - 7 = 4$

21 $7 = 2z + 6$ **27** $3x + 1 = 11$ **33** $9 = 6a - 27$

22 $6x + 1 = -5$ **28** $2x + 4 = 14$ **34** $7 = 1 - 2x$

23 $5 = 7x - 23$ **29** $16 = 7x - 1$ **35** $10 + 2x = -2$

24 $5x + 1 = 4.5$ **30** $2.4 = 3x - 1.2$ **36** $2 + 3x = 7.4$

I think of a number, double it and add 3. The result is 15. What is the number?

Let the number be x $2x + 3 = 15$
Take 3 from both sides $2x = 12$
Divide both sides by 2 $x = 6$
The number is 6.

In the remaining problems in this exercise form an equation and solve it.

37 I think of a number, multiply it by 4 and subtract 8.
The result is 20.
What was the number?

38 I think of a number, multiply it by 3 and add 6. The result is 21.
What is the number?

39 I think of a number, multiply it by 3 and add the result to 7. The total is 28.
What is the number?

40 The sides of a rectangle are x cm and 3 cm.
The perimeter is 24 cm.
Find x.

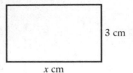

3 cm

x cm

41 The lengths of the three sides of a triangle are x cm, x cm and 6 cm.
The perimeter is 20 cm.
Find x.

42 Mary and Jean each have x sweets and Susan has 10 sweets. They have 24 sweets altogether.
What is x?

43 Three boys had x sweets each. They gave 9 sweets to a fourth boy and then found that they were left with 18 sweets among the three of them.
Find x.

44 I have two pieces of ribbon each x cm long and a third piece 9 cm long. Altogether there are 31 cm of ribbon.
What is the length of each of the first two pieces?

45 Jen is y years old and her mother is 27 years older. Together their ages total 45 years.

a In terms of Jen's age, how old is her mother now?

b Form an equation in y and solve it to find Jen's age and her mother's age now.

46 A fishing rod is 27 ft long and consists of three parts. The first part is x ft long, the second part is 1 ft longer than the first and the third part is 1 ft longer than the second part.
Form an equation in x and solve it to find the length of each part.

47 Sonia is n years old now and Cynthia, her sister, is 2 years older. In 5 years time the sum of their ages will be 20.
How old is Sonia now?
How old will Cynthia be in 5 years time?

SIMPLIFYING
EXPRESSIONS

We can shorten, i.e. *simplify*, $2x + x$ to $3x$
because $2x$ means $x + x$,

i.e. $2x + x = x + x + x$
$= 3x$

Like Terms

Consider $3x + 5x - 4x + 2x$.

This is called an *expression* and can be simplified to $6x$.

$3x, 5x, 4x$ and $2x$ are all *terms* in this expression. Each term contains x.
They are of the same type and are called *like terms*.

Like terms can be simplified using the ordinary rules of addition and
subtraction.

EXERCISE 21H

Simplify $4h - 6h + 7h - h$

Remember the sign in front of a term tells
you what to do with just that term, so

$4h - 6h + 7h - h = 4h + 7h - 6h - h$
$= 11h - 7h$
$= 4h$

$4h - 6h + 7h - h = 4h$

Simplify

1 $3x + x + 4x + 2x$

6 $9y - 3y + 2y$

2 $3x - x + 4x - 2x$

7 $2 - 3 + 9 - 1$

3 $-6x + 8x$

8 $-16 - 3 - 4$

4 $6x - 1 + 4 - 7$

9 $-3x + 5x - 1$

5 $-8x + 6x$

10 $-2x - x + 3x$

Unlike Terms

$3x + 2x - 7$ can be simplified to $5x - 7$,
and $5x - 2y + 4x - 3y$ can be simplified to $9x - 5y$.

Terms containing x are different from terms without an x.
They are called *unlike terms* and cannot be collected.

Similarly $9x$ and $5y$ are unlike terms; therefore $9x - 5y$ cannot be
simplified.

Simplify   **a** $3x + 4 - 7 - 2x + 4x$ **b** $2x + 4y - x - 5y$

a $3x + 4 - 7 - 2x + 4x = 5x - 3$

$3x + 4 - 7 - 2x + 4x = 3x - 2x + 4x + 4 - 7$
$$= 5x - 3$$
$5x$ and 3 are unlike terms,
so $5x - 3$ cannot be simplified

b $2x + 4y - x - 5y = x - y$

x and y are unlike terms
so $x - y$ cannot be simplified

Simplify

1 $2x + 4 + 3 + 5x$ **4** $4a + 5c - 6a$

2 $2x - 4 + 3x + 9$ **5** $6x - 5y + 2x + 3y$

3 $5x - 2 - 3 - x$ **6** $6x + 5y + 2x + 3y$

7 $6x + 5y + 2x - 3y$ **12** $9x + 3y - 10x$

8 $6x + 5y - 2x + 3y$ **13** $2x - 6y - 8x$

9 $4x + 1 + 3x + 2 + x$ **14** $7 - x - 6 - 3x$

10 $6x - 9 + 2x + 1$ **15** $8 - 1 - 7x + 2x$

11 $7x - 3 - 9 - 4x$ **16** $9x - 1 + 4 - 11x$

17 $6x - 5y + 2x + 3y + 2x$ **20** $-2z + 3x - 4y + 6z + x - 3y$

18 $6x - 5y - 2x - 3y + 7x - y$ **21** $4x + 3y - 4 + 6x - 2y - 7 - x$

19 $30x + 2 - 15x - 6 + 4$ **22** $7x + 3 - 9 - 9x + 2x - 6 + 11$

EQUATIONS WITH LETTER TERMS ON BOTH SIDES

Some equations have letter terms on both sides. Consider the equation

$$5x + 1 = 2x + 9$$

We want to have a letter term on one side only so we need to take $2x$ from both sides. This gives

$$3x + 1 = 9$$

and we can go on to solve the equation as before.

Notice that we want the letter term on the side which has the greater number of xs to start with.

If we look at the equation

$$9 - 4x = 2x + 4$$

we can see that xs have been taken away on the left-hand side, so there are more xs on the right-hand side.
Add $4x$ to both sides and then the equation becomes

$$9 = 6x + 4$$

and we can go on as before.

EXERCISE 21J

Deal with the letters first, then the numbers.

Solve $5x + 2 = 2x + 9$

$$5x + 2 = 2x + 9$$

Deal with the x terms first.

Take $2x$ from both sides	$3x + 2 = 9$
Take 2 from both sides	$3x = 7$
Divide both sides by 3	$x = \frac{7}{3} = 2\frac{1}{3}$

Solve the equations.

1 $3x + 4 = 2x + 8$

2 $x + 7 = 4x + 4$

3 $2x + 5 = 5x - 4$

4 $3x - 1 = 5x - 11$

5 $7x + 3 = 3x + 31$

6 $6z + 4 = 2z + 1$

7 $7x - 25 = 3x - 1$

8 $11x - 6 = 8x + 9$

Solve $9 + x = 4 - 4x$

$$9 + x = 4 - 4x$$

> The left-hand side contains the greater number of xs so we add $4x$ to both sides to remove the xs from the right-hand side.

Add $4x$ to both sides $9 + 5x = 4$

Take 9 from both sides $5x = -5$

Divide both sides by 5 $x = -1$

Check: If $x = -1$, left-hand side $= 9 + (-1)$

$$= 8$$

right-hand side $= 4 - (-4)$

$$= 8$$

So $x = -1$ is the solution.

Solve the equations.

9 $4x - 3 = 39 - 2x$ **13** $5x - 6 = 3 - 4x$

10 $5 + x = 17 - 5x$ **14** $12 + 2x = 24 - 4x$

11 $7 - 2x = 4 + x$ **15** $32 - 6x = 8 + 2x$

12 $24 - 2x = 5x + 3$ **16** $9 - 3x = -5 + 4x$

EQUATIONS CONTAINING LIKE TERMS

If there are a lot of terms in an equation, first collect the like terms on each side separately.

EXERCISE 21K

Solve $2x + 3 - x + 5 = 3x + 4x - 6$

$$2x + 3 - x + 5 = 3x + 4x - 6$$
$$x + 8 = 7x - 6$$

> Simplify each side.

Take x from both sides $8 = 6x - 6$

Add 6 to both sides $14 = 6x$

Divide both sides by 6 $\frac{14}{6} = x$

$$x = \frac{7}{3} = 2\frac{1}{3}$$

Solve the following equations.

1 $3x + 2 + 2x = 7$ **6** $3x + 2x - 4x = 6$

2 $7 + 3x - 6 = 4$ **7** $7 = 2 - 3 + 4x$

3 $6 = 5x + 2 - 4x$ **8** $5x + x - 6x + 2x = 9$

4 $9 + 4 = 3x + 4x$ **9** $5 + x - 4x + x = 1$

5 $3x + 8 - 5x = 2$ **10** $6x = x + 2 - 7 - 1$

11 $5x + 6 + 3x = 10$ **14** $1 - 4 - 3 + 2x = 3x$

12 $8 = 7 - 11 + 6x$ **15** $3x - 4x - x = x - 6$

13 $7 + 2x = 12x - 7x + 2$ **16** $2 - 4x - x = x + 8$

Solve $\quad 9 - 3x = 15 - 4x$

$$9 - 3x = 15 - 4x$$

Notice that there is a greater deficit of xs on the right so we get rid of the xs on that side.

Add $4x$ to both sides $\quad 9 + x = 15$

Take 9 from both sides $\quad\quad x = 6$

---

17 $5 - 3x = 1 - x$ **21** $16 - 6x = 1 - x$

18 $16 - 2x = 19 - 5x$ **22** $4 - 3x = 1 - 4x$

19 $6 - x = 12 - 2x$ **23** $4 - 2x = 8 - 5x$

20 $-2 - 4x = 6 - 2x$ **24** $3 - x = 5 - 3x$

25 $6 - 3x = 4x - 1$ **29** $13 - 4x = 4x - 3$

26 $4z + 1 = 6z - 3$ **30** $7x + 6 = x - 6$

27 $3 - 6x = 6x - 3$ **31** $6 - 2x = 9 - 5x$

28 $8 - 4x = 14 - 7x$ **32** $3 - 2x = 3 + x$

Solve $3 - 2x = 5$

$$3 - 2x = 5$$

> The left-hand side has a deficit of x s so we collect them on the right.

Add $2x$ to both sides $3 = 5 + 2x$
Take 5 from both sides $-2 = 2x$
Divide both sides by 2 $x = -1$

33 $13 - 4x = 5$ **35** $6 = 8 - 3x$

34 $6 = 2 - 2x$ **36** $0 = 6 - 2x$

37 $9x + 4 = 3x + 1$ **42** $5 - 3x = 2$

38 $2x + 3 = 12x$ **43** $6 + 3x = 7 - x$

39 $7 - 2x = 3 - 6x$ **44** $5 - 2x = 4x - 7$

40 $3x - 6 = 6 - x$ **45** $5x + 3 = -7 - x$

41 $-4x - 5 = -2x - 10$ **46** $4 - 3x = 0$

47 $-2x + x = 3x - 12$ **54** $4 - x - 2 - x = x$

48 $-4 + x - 2 - x = x$ **55** $4 - x - 2 + x = x$

49 $3x + 1 + 2x = 6$ **56** $2x + 7 - 4x + 1 = 4$

50 $4x - 2 + 6x - 4 = 64$ **57** $6 - 3x - 5x - 1 = 10$

51 $2x + 7 - x + 3 = 6x$ **58** $6x + 3 + 6 = x - 4 - 2$

52 $6 - 2x - 4 + 5x = 17$ **59** $x - 3 + 7x + 9 = 10$

53 $9x - 6 - x - 2 = 0$ **60** $15x + 2x - 6x - 9x = 20$

61 An operator on a production line in a car assembly plant is given a box containing n windscreen wipers. She uses 3 to a car until all that remain are 2 faulty wipers. A check shows that she has fitted the wipers on 58 cars.
Form an equation in n and solve it.
How many wipers were in the box to start with?

62 Nia weighs 4 kg less than Madge who weighs 3 kg less than Penny. Altogether they weigh 137 kg.

If Madge weighs x kg, what is the weight of each of Penny and Mia in terms of x?

How much does Madge weigh?

63 Laura goes to a shop that sells two models of radio. One cost three times as much as the other and the two together cost £72. Laura's mother wants to know the cost of the cheaper one.

If the cheaper radio costs £x form an equation in x.

What should Laura's answer be?

64 Keith and Sheila go into a cafe because they want a hot drink. Keith has a cup of tea and a cake, while Sheila has a cup of coffee and 2 cakes. Coffee costs 15 p more than tea, and each cake is 35 p more than a cup of coffee.

a If a cup of tea costs x pence write down, in terms of x, the cost of

   i a cup of coffee   **ii** a cake   **iii** 2 cakes.

b The total cost of the tea, coffee and cakes is £4.15.

Form an equation in x and solve it.

Hence write down the cost of

   i a cup of tea   **ii** a cup of coffee   **iii** a cake.

65 Divide 45 into two parts so that if 4 is subtracted from the larger number the result is 7 more than the smaller number.

66 Four brothers play in a cricket team. In their last match Jim scored 15 more runs than Norman, Dennis scored five times as many as Norman and sixteen more than Pete. In total they scored 107 runs. How many did each brother score?

67 Two sisters, Janet and Nora, each have a box holding 20 chocolates. Janet eats 5 and gives some away to her friends. Nora gives 1 away and eats three times as many as Janet has given away. When they compare boxes, each sister still has the same number of chocolates left as the other.

If Janet gave x chocolates away, form an equation in x and solve it. How many chocolates did Nora eat?

1 A cola costs 10 p more than an orange squash.
If the orange squash costs x pence and the cola costs 75 pence, form an equation in x and solve it.
How much does an orange squash cost?

2 Solve $5y = 45$

3 Solve the equation $5x - 0.7 = 2.8$

4 Solve the equation $3x + 2 = 4$

5 I think of a number, add 4 and the result is 10.
Form an equation and solve it to find the number I thought of.

6 Solve the equation $6x + 2 = 3x + 8$

7 Solve the equation $4x - 2 = -6$

8 Simplify $4x - 3y + 5x + 2y$

9 Solve the equation $4x + 2 - x = 6$

10 Solve the equation $14 - 3x = 5$

1 An iced slice costs 14 pence less than a cream doughnut.
If a cream doughnut costs n pence and an iced slice costs 27 pence, form an equation in n and solve it.
How much does a cream doughnut cost?

2 Solve $5x = 1$

3 Solve the equation $4x = 0.5$

4 Solve the equation $4x - 5 = 3$

5 Simplify $3c - 5c + 9c$

6 Solve the equation $3x - 2 = 4 - x$

7 When I think of a number, double it and add three, I get 11.
What number did I think of?

8 Solve the equation $x + 2x - 4 = 9$

9 Simplify $2a + 4 - 3 + 5a - a$

10 Solve the equation $12 - x = 6 - 2x$

INVESTIGATION

Meg wanted to find out Malcolm's age without asking him directly what it was.

The following conversation took place.

Meg: Think of your age but don't tell me what it is
Malcolm: Right
Meg: Multiply it by 5, add 4 and take away your age.
Malcolm: Yes
Meg: Divide the result by 4 and tell me your answer.
Malcolm: 15
Meg: That means you are 14.
Malcolm: Correct. How do you know that?

However many times Meg tried this on her friends and relations she found their age by taking 1 away from the number they gave.

Does it always work?

Can you use simple algebra to prove that it always gives the correct answer?

GROUPING DATA

This list shows how much, in pence, each pupil in Class 7H spent in the school canteen one day.

25	35	48	52	62	63	80	95	97	97
101	103	108	115	132	132	134	139	142	146
150	151	154	163	170	179	185	194	210	217

Any number from 0 to about 250 could appear in a list like this one. In this list, a couple of numbers appear twice but most of the possible numbers do not appear at all.

- If this information has been collected to find the total amount spent, then each individual amount is needed. If we attempted to collect amounts like these on a tally chart, we would have to list every number from 0 to about 250. Doing this would take a long time, so collecting them as a simple list and putting them in order later would be sensible.

- On the other hand, if we want some idea of the spending pattern, the detail given by having each individual amount is not important; a clearer picture of the pattern can be obtained by collecting the amounts in groups of, say, less than 50 p, 50 p to less than £1, and so on. We can then collect the figures on a tally chart like this one.

Amount in pence	Frequency
0 to 49	///
50 to 99	//// //
100 to 149	//// ////
150 to 199	//// ///
200 to 249	//

Notice that, without the detail of the exact figures, we would not know whether any one had spent exactly £1, for instance.

1 The following questions need answering about the marks in last summer's maths exam for year 7 pupils.

   a How many children got between 50% and 59%?

   b Did anyone get exactly 50%?

   c To the nearest 10, what was the commonest mark?

   d How many children got less than 25%?

There is some information available; for each of these questions, discuss whether you need to know

- the range of the marks
- the number of possible marks
- each individual mark
- grouped marks.

2 Repeat question **1** for these questions which refer to the numbers of people queuing at a supermarket checkout.

a How often was the queue less than five people?

b How many times were there two people in the queue?

c What was the largest number recorded?

3 Repeat question **1** for these questions which refer to the prices of personal cassette players.

a I have £25 to spend. Can I buy a player for this?

b How many different models are available for between £20 and £30?

c What proportion of the models cost less than about £40?

d In what price range is there the largest choice of models?

DISPLAYING GROUPED DATA IN A BAR CHART

When information has been collected using a grouped tally chart, we can illustrate it using a bar chart. The data, given opposite, about the sum of money spent in the school canteen can be shown in this bar chart.

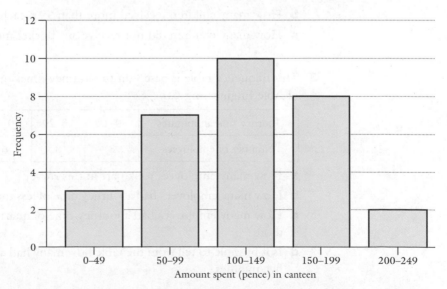

EXERCISE 22B

1 For an English project pupils were asked to investigate the number of letters per word from a paragraph of the book *Tom's Midnight Garden*. This is the list of the number of letters per word.

```
4  3  3  5  3  12  4  6  4  3  9  2  4   4   1  4
6  4  3  6  4   6  3  7  4  1  6  4  5   3   4  3
6  4  3  4  6   6  5  3  3  2  7  3  4  10  11
```

a How many words were there in the paragraph?

b Copy and complete this frequency table.

Number of letters	Tally	Frequency
1–3		
4–6		
7–9		
10–12		
	Total:	

c Draw a bar chart to illustrate the information in this frequency table.

2 A group of ten-year-olds were asked how much pocket money they were given each week. This frequency table was made from the information.

Weekly pocket money	0–49	50–99	100–149	150–199	200–249
Number of children	10	15	42	68	18

a How many children were asked how much pocket money they received?

b How many children received more than 99 p each week?

c How many children did not receive any pocket money?

3 This frequency table is based on the journey times of the employees of Able Engineering Co.

Journey time in minutes	0–14	15–29	30–44	45–59
Number of employees	3	8	6	2

a How many employees took part in this survey?

b How many employees had a journey time of less than half an hour?

c How many employees had a journey time of quarter of an hour or more?

d Is it possible to tell from the table how many had a journey time of five minutes?

4 Investigate the number of letters per word in question **5** below.

    a Make a list of the number of letters per word.

    b Make a frequency table for this list using the groups 1–3 letters, 4–6 letters, 7–9 letters, 10–12 letters, more than 12 letters.

    c Draw a bar chart to illustrate the information in your frequency table.

5 Make your own survey of the pocket money given to your friends and classmates.

    a Start by preparing a frequency table like the one in question **2**, but leave the numbers of children blank.
You may have to add further groups: 250–299, and so on.

    b Ask about 10 children to put a tally mark in the appropriate column of your table. (Some of your friends may not want to tell you how much money they are given, and asking them to draw the tally mark avoids possible embarrassment. Do not press anyone who is reluctant to give the information.)

    c Illustrate your survey with a bar chart.

    d Compare your bar chart with those produced by other members of your class.

6 This bar chart illustrates a survey into the number of books (not counting school books) read each week by some sixteen-year-olds.

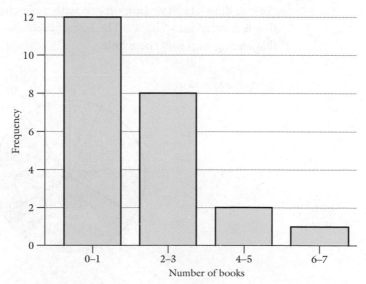

    a How many read less than two books each week?

    b How many read four or more books each week?

    c Why is it not possible to tell how many read four books each week?

PIE CHARTS

Pie charts are used to show the fraction that each category or group is of the whole.

This is the frequency table showing the sums of money spent in the school canteen each day by 30 pupils.

Amount in pence	Frequency
0 to 49	3
50 to 99	7
100 to 149	10
150 to 199	8
200 to 249	2
	Total 30

If we are interested in the fraction that the number in each group is of the 30 pupils, we can write these fractions down from the information in the table,

i.e. $\frac{3}{30}$ of the group spent 0 to 49 p,

$\frac{7}{30}$ of the group spent 50 p to 99 p, and so on.

We can illustrate these proportions with a pie chart; a pie chart is a circle that is divided into 'slices'. The size of each slice represents the fraction that each group is of the whole.

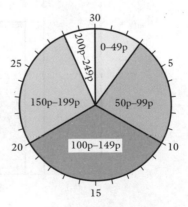

Pie charts do not usually come with the circumference of the circle divided into a convenient number of parts. Each slice is usually labelled with what it represents and its percentage of the whole quantity.

This pie chart is taken from *Social Trends 22* and shows average household water use.

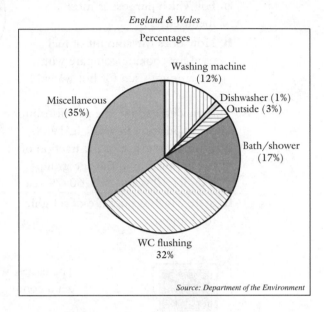

England & Wales

Percentages

Washing machine (12%)

Dishwasher (1%)
Outside (3%)

Miscellaneous (35%)

Bath/shower (17%)

WC flushing 32%

Source: Department of the Environment

Without looking at the percentages we can see that the proportion of water used for WC flushing is about $\frac{1}{3}$ of the total used and the water used for washing machines is about a third that used for WC flushing, i.e. about $\frac{1}{9}$ of the total water used.

EXERCISE 22C

1 This pie chart shows the uses made of personal computers.

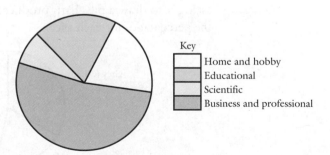

Key

Home and hobby
Educational
Scientific
Business and professional

a For which purpose were computers used most?

b Estimate the fraction of the total numbers used for

i scientific purposes   **ii** home and hobbies

2 The pie chart opposite shows how fuel is used for different purposes in the average house:

a For which purpose is most fuel used?

b How does the amount of fuel used for cooking compare with the amount used for hot water?

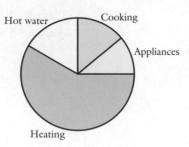

3 This pie chart shows the age distribution of the population in years, in 1988.

a Estimate the size of the fraction of the population in the age groups

    i under 10 years   **i** 60–79 years.

b State which groups are of roughly the same size.

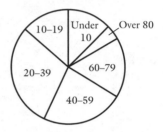

DRAWING PIE CHARTS

The easiest way to draw a pie chart is to get a computer to do it for you.

This table shows the numbers of people with eyes of certain colours.

Eye colour	Brown	Hazel	Blue	Grey
Frequency	22	6	12	20

Entering the last four columns into a spreadsheet program, and then asking it to draw a pie chart, produced this diagram – it even works out the percentages for each slice.

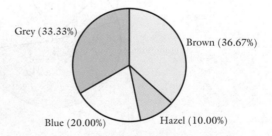

To draw it ourselves, we have to work out the size of each pie slice; this is given by the size of angle at the 'point', i.e. at the centre. Therefore we need to calculate the sizes of the angles.

The number of people is 60, and $\frac{1}{60}$ of 360° is 6°.

As there are 12 blue-eyed people, they form $\frac{12}{60}$ of the whole group and are therefore represented by that fraction of the circle.

Blue: $\frac{12}{60}$ of 360° = 72°   Grey: $\frac{20}{60}$ of 360° = 120°

Hazel: $\frac{6}{60}$ of 360° = 36°   Brown: $\frac{22}{60}$ of 360° = 132°

(Check: total of the angles is 360°)

Now draw a circle of radius about 5 cm
(or whatever is suitable). Draw one radius as
shown and complete the diagram using a
protractor, turning your page into the easiest
position for drawing each new angle.
Label each 'slice'.

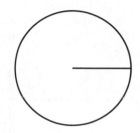

EXERCISE 22D

Draw pie charts to represent the following information, first working out the angles.

1 A box of 60 coloured balloons contains the following numbers of balloons of each colour.

Colour	Red	Yellow	Green	Blue	White
Number of balloons	16	22	10	7	5

2 Ninety people were asked how they travelled to work and the following information was recorded.

Transport	Car	Bus	Train	Motorcycle	Bicycle
Number of people	32	38	12	6	2

3 On a cornflakes packet the composition of 120 g of cornflakes is given in grams as follows.

Protein	Fat	Carbohydrate	Other ingredients
101	1	10	8

4 Of 90 cars passing a survey point it was recorded that 21 had two doors, 51 had four doors, 12 had three (two side doors and a hatchback) and 6 had five doors.

5 A large flower arrangement contained 18 dark red roses, 6 pale pink roses, 10 white roses and 11 deep pink roses.

6 The children in a class were asked what pets they owned and the following information was recorded.

Animal	Dog	Cat	Bird	Small animal	Fish
Frequency	8	10	3	6	3

Sometimes the total number involved does not work as conveniently as in the previous problems. In this case we find the angle correct to the nearest degree.

The eye colour of 54 people is recorded in the table. Draw a pie chart to show the proportions of this group with each different eye colour.

Eye colour	Blue	Grey	Hazel	Brown	
Frequency	10	19	5	20	Total: 54

Angles: Blue: $\frac{10}{54}$ of $360° = 66.6\ldots°$ $\boxed{\begin{array}{l}\text{Use your calculator to find}\\ (\,360 \div 54\,) \times 10\end{array}}$

$= 67°$ to the nearest degree

Grey: $\frac{19}{54}$ of $360° = 126.6\ldots°$

$= 127°$ to the nearest degree

Hazel: $\frac{5}{54}$ of $360° = 33.3\ldots°$

$= 33°$ to the nearest degree

Brown: $\frac{20}{54}$ of $360° = 133.3\ldots°$

$= 133°$ to the nearest degree

Check: $67 + 127 + 33 + 133 = 360$

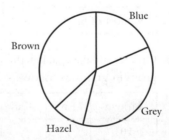

In this case, adding up the angles gives $360°$.

Sometimes, because of rounding, the sum of the calculated angles comes to one or two degrees more or less than $360°$. If the sum is more than $360°$, find the rounded-up angle that is nearest to half way between two values, and round it down. If the sum is less than $360°$, find the rounded-down angle that is nearest to half way between two values and round it up.

Draw pie charts to represent the following information. Work out the angles first and, where necessary, give the angles to the nearest degree. Try using a chart drawing package for some of these.

7 300 people were asked whether they lived in a flat, a house, a bedsit, a bungalow or in some other type of accommodation and the following information was recorded.

Type of accommodation	Flat	House	Bedsit	Bungalow	Other
Frequency	90	150	33	15	12

8 In a street in which 80 people live the numbers in various age groups are as follows.

Age group (years)	0–15	16–21	22–34	35–49	50–64	65 and over
Number of people	16	3	19	21	12	9

9 A group of people were asked to select their favourite colour from a card showing 6 colours and the following results were recorded.

Colour	Rose pink	Sky blue	Golden yellow	Violet	Lime green	Tomato red
Number of people (frequency)	6	8	8	2	1	10

10 Peter recorded the types of vehicle moving along a road during one hour and drew up this table.

Vehicle	Cars	Vans	Lorries	Motorcycles	Bicycle
Frequency	62	11	15	10	2

11 This table showing the hours of sunshine per day during May was compiled using information from the school's weather station.

Hours per day	0–2	3–4	5–6	7–8	over 8
Frequency	3	9	11	5	2

12 This table, from *Social Trends 22*, shows average attendances at football matches for various years.

	Football League (England & Wales)			
	Division 1	Division 2	Division 3	Division 4
1961/62	26,106	16,132	9,419	6,060
1966/67	30,829	15,701	8,009	5,407
1971/72	31,352	14,652	8,510	4,981
1976/77	29,540	13,529	7,522	3,863
1980/81	24,660	11,202	6,590	3,082
1986/87	19,800	9,000	4,300	3,100
1987/88	19,300	10,600	5,000	3,200
1988/89	20,600	10,600	5,500	3,200
1989/90	20,800	12,500	5,000	3,400
1990/91	22,681	11,457	5,208	3,253

1 Football league attendances are rounded to the nearest hundred between 1986/87 and 1989/90.   Source: Football League

a Draw a pie chart showing the proportions attending each division in 1961/62.

b Draw another pie chart showing the proportions attending each division in 1990/91.

c Have the proportions changed between the 1961/62 season and the 1990/91 seasons?

13 This chart shows the percentage of spending money used on various categories of purchases by a group of 12-year-olds.

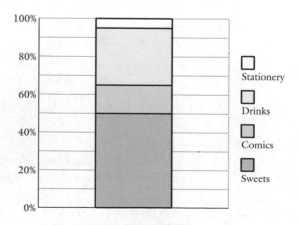

Show this information in a pie chart.

PRACTICAL
WORK

1 This is a group activity for the whole class.
Think of a question about the money that the children in your group spend; then gather and organise the information so that you can attempt to answer some questions.

Some questions you might like to answer are

- How much do my class-mates spend on fares each day?
- How much money do my class-mates have each week?
- What do they spend it on?
- How much of that money are they given and how much do they have to earn by doing tasks around the home?
- How does the spending pattern in the school canteen by my group compare with the data given at the beginning of this chapter?

When you have decided what question you want to answer, think about the information that you need to collect and how you are going to collect it.
Will you group the information – if so what groups will you use?
When you have organised the information, what does it show?
What is the best way to illustrate the information?

2 This is an individual activity.
Choose two different daily newspapers. For each one find the area of the front page used by each of the categories HEADLINES, PICTURES, TEXT, EVERYTHING ELSE.
Illustrate your results with either bar charts or pie charts and give a reason for your choice.
What comparisons can you make between the two front pages?

SUMMARY 5

SUMMARISING DATA

For a list of values,

- the *range* is the difference between the largest value and the smallest value,
- the *mean* is the sum of all the values divided by the number of values,
- the *median* is the middle value when they have been arranged in order of size, (when the middle of the list is half way between two values, the median is the average of these two values),
- the *mode* is the value that occurs most frequently.

PIE CHARTS

Pie charts are used to show the fraction that each group or category is of the whole list.

The size of a 'slice' is given by the angle at the centre of the circle. This angle is found by first finding the number of values in the group as a fraction of all the values, and then finding this fraction of $360°$.

VOLUME

Volume is measured by standard sized cubes.

A cube with edges 1 cm long has a volume of 1 cubic centimetre, written $1 \, cm^3$.

$$1 \, cm^3 = 10 \times 10 \times 10 \, mm^3 = 1000 \, mm^3$$

$$1 \, m^3 = 100 \times 100 \times 100 \, cm^3 = 1\,000\,000 \, cm^3$$

CAPACITY

The capacity of a container is the volume of liquid it will hold.

The main metric units of capacity are the litre (1) and the millilitre (ml), where

$$1 \text{ litre} = 1000 \, ml, \quad 1 \text{ litre} = 1000 \, cm^3 \quad \text{and} \quad 1 \, ml = 1 \, cm^3$$

The main imperial units of capacity are the gallon and the pint, where

$$1 \text{ gallon} = 8 \text{ pints}$$

Rough conversions between metric and imperial units of capacity are given by

$$1 \text{ litre} \approx 1.75 \text{ pints} \quad \text{and} \quad 1 \text{ gallon} \approx 4.5 \text{ litres}$$

**ALGEBRAIC
EXPRESSIONS**

Terms such as $2x$ and $5x$ mean $2 \times x$, i.e. $x + x$

and $5 \times x$, i.e. $x + x + x + x + x$

Similarly ab means $a \times b$

$2x + 5x$ can be simplified to $7x$, since

$2x + 5x = x + x \; + \; x + x + x + x + x$

$\qquad\qquad = 7x$

$2a + 3b$ cannot be simplified.

**SOLVING
EQUATIONS**

An equation is a relationship between an unknown number, represented by a letter, and other numbers, e.g. $2x - 3 = 5$

Solving the equation means finding the unknown number,
i.e. ending up with $x = (\text{a number})$

Provided that we do the same to both sides of an equation, we keep the equality; this can be used to solve the equation,

e.g.   to solve $\qquad\qquad\qquad 2x - 3 = 5,$

First add 3 to both sides $\qquad 2x - 3 + 3 = 5 + 3$

This gives $\qquad\qquad\qquad\qquad 2x = 8$

Now divide each side by 2 $\qquad\quad x = 4$

**REVISION
EXERCISE 5.1
(Chapters 19 & 20)**

1 In three different bookshops a particular school textbook costs £12.49, £14.49 and £14.99.

a What is the mean price?

b What is the range of prices?

2 Find the mode and the median of the numbers

$$5, 4, 7, 4, 5, 9, 8, 7, 4, 3, 5, 4, 8.$$

3 Tim and Meg sold bunches of flowers which they had put together. Out of curiosity they counted the number of flowers in several bunches to see how the numbers that Tim put in a bunch compared with the numbers that Meg put in a bunch. The numbers in Tim's bunches were

$$12, 17, 11, 15, 20, 11, 16, 19, 21, 8$$

and the numbers in Meg's bunches were

$$17, 15, 14, 16, 13, 12, 15, 15, 16, 17$$

Compare the mean and range of the number of flowers in Tim's bunches with the mean and range of the number of flowers in Meg's bunches.

4 Tara chose four cards at random several times from an ordinary pack of 52 playing cards. Each time, she counted the number of the cards that were red. Her results are given in the table.

Number of red cards	Frequency
0	1
1	5
2	9
3	7
4	3

a How many times did she choose 4 cards?

b What was the median number of red cards she chose?

c How many red cards did she choose altogether?

d What was the modal number of red cards she chose?

e What was the mean number of cards she chose?

5 Use isometric paper to draw a cube of side 4.5 cm.

6 This presentation box for holding a watch is 6 cm long, 4 cm wide and 2 cm high.

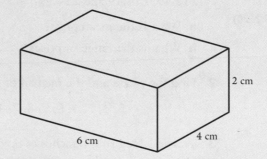

a How many faces does it have?

b Sketch the faces, showing their measurements.

c On 1 cm grid paper, draw a net that will make this box.

7 Express   **a** $9\,cm^3$ in mm^3 **b** $2\,m^3$ in cm^3

8 Give, roughly

a the number of pints in 5 litres

b the number of litres in 20 pints.

9 **a** Find the volume of a solid wooden cube of side 2 cm.

  b How many cubes of side 2 cm will fit inside a cubical box of side 10 cm?

10 Petrol costs 60 p a litre.
Roughly, how much is this a gallon?

REVISION
EXERCISE 5.2
(Chapters 21 & 22)

1 Solve the equations

  a $a + 4 = 11$ **c** $3.5 + x = 8$

  b $b - 3 = 12$ **d** $y - 3.5 = 6$

2 Solve the equations

  a $6z = 24$ **b** $3x + 2 = 17$

3 Simplify

  a $5x - 4x - 7x$ **b** $5a - 4b - 3a + 2b$

4 Solve the equations

  a $5x - 2x = x + 8$ **b** $7x - 6 - 2x - 14 = 0$

5 Pete is p years old now and his sister Carol is 3 years older.
In 10 years time the sum of their ages will be 33.
How old is Pete now?
How old will Carol be in 10 years time?

6 I think of a number, add 3 and multiply the answer by 4. The result is 40.
What is the number?

7 An hotel group grades its hotels from 1 star (*) up to 5 star (*****) depending on the facilities the hotel offers. The number of hotels in each grade is given in the table.

Grade of hotel	Number of hotels in grade
1 star	2
2 star	6
3 star	11
4 star	12
5 star	5

  a How many hotels are there in the group?

  b Draw a bar chart to illustrate the information in this frequency table.

8 This frequency table gives the time taken by the pupils in a group to solve a puzzle.

Time in minutes	0–4	5–9	10–14	15–19
Number of pupils	3	10	7	2

a How many pupils were there in the group?

b How many pupils took less than 10 minutes?

c How many pupils took 10 minutes or more?

d Can you tell how many pupils took 10 minutes?

9 The pie chart shows the choice of juices taken for breakfast by the 60 guests in a hotel one morning.

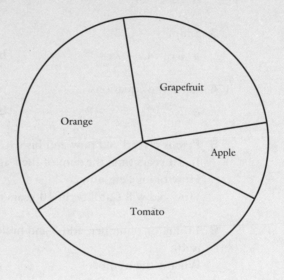

Estimate the sizes of the angles and hence find, approximately, the number who preferred

a orange juice   **b** tomato juice   **c** apple juice.

10 A survey of Year 7 pupils gave the following information about the way they came to school.

Way of coming to school	Bus	Car	Walk	Bicycle
Number of pupils	43	21	14	12

a How many pupils in Year 7 took part in the survey?

b Draw a pie chart to represent this information after you have calculated the angles corresponding to each slice.

**REVISION
EXERCISE 5.3
(Chapters 19 to 22)**

1 The pie chart shows how 40 chocolates are shared among three sisters.

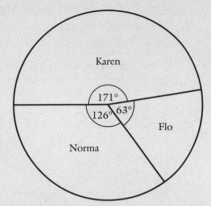

a Find each angle as a fraction of a revolution.

b How many chocolates did Karen get?

c How many did Norma get?

d How many did Flo get?

2 Use isometric paper to draw a cuboid measuring 3 cm by 5 cm.

3 This net will make a cube.

a Which edge meets
  i AB   **ii** FG?

b Which other corners meet at
  i J   **ii** G?

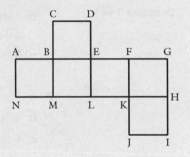

4 Find the capacity, in litres, of a rectangular metal tank measuring 40 cm by 30 cm by 1 m.

5 Express   **a** $5000 \, \text{mm}^3$ in cm^3 **b** $0.5 \, \text{m}^3$ in cm^3

6 Solve the equations
  a $p + 6 = 9$ **b** $q - 10 = 12$

7 I think of a number, subtract 4 and multiply by 6. The answer is 48. What is the number?

8 Solve the equation $2x + 7 = 5x - 14$

9 Simplify
  a $5x - 3 + 4x - 1$ **b** $3x + 5y - 2x + 4y$

10 Once every ten minutes Diana counted the number of people standing at the bus stop which was outside her home. Her results are given in the table.

Number of people at bus stop	Frequency
0	6
1	2
2	3
3	6
4	2
5	1

a How many times did she count?

b How long did she spend gathering the information?

c What was the mean number of people waiting at the bus stop?

REVISION EXERCISE 5.4 (Chapters 1 to 22)

1 a Find the difference between the value of the 4 in 3408 and the 7 in the number 1079.

b Estimate the value of $742 + (53 - 37) \times 22$ and then use your calculator to find its exact value.

c What is the remainder when 659 is divided by 56?

2 a Find **i** $4\frac{5}{8} + 7\frac{4}{5}$ **ii** $18\frac{2}{7} - 6\frac{7}{12}$

b Which is the greater: $\frac{7}{21}$ or $\frac{5}{14}$?

c Change

i 0.86 into a percentage

ii 34% into a decimal

iii $\frac{27}{50}$ into a percentage.

3 a Find, without using a calculator

i 0.043×10 **ii** $66 \div 100$ **iii** 0.057×30 **iv** $1.28 \div 40$

b From 3.74 subtract the difference between 4.53 and 5.27

4 a Estimate the value of **i** 63×382 **ii** $6394 \div 278$.
Now use your calculator to find the exact values.

b Use your calculator to change the fractions
$\frac{4}{5}, \frac{19}{25}, \frac{17}{20}, \frac{2}{3}$ and $\frac{5}{8}$ into decimals.
Use these values to put the fractions in order of size with the smallest first.

5 Express

  a 0.46 cm in mm

  b 0.55 km in m

  c 240 mm^2 in cm^2

  d 0.0008 m^3 in cm^3

  e 7690 kg in tonnes

  f 750 000 cm^3 in m^3

6 Find the size of each angle marked with a letter.

  a 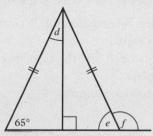 **b**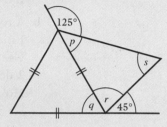

7 Write down the next three numbers in each sequence.

  a 8, 5, 2,   **b** −12, −9, −6

8 a The formula for finding profit £p when n boxes of chocolates are
    sold is $p = 2.5n - 30$
    Find p when $n = 10$ and explain what you think that answer
    means.

  b I think of a number, double it and then add 15 to the result. The
    answer is three times the number I thought of.
    If x stands for the number I first thought of, form an equation in
    x and solve it.

9 One side of a rectangle is 5 cm longer than another. Its perimeter
    is 50 cm.
    What is the length of **a** a long side   **b** a short side?

10 Find the mode and median of the following set of numbers.

    4.1, 3.8, 4.2, 4.3, 3.9, 3.8, 4.5, 3.7

**REVISION
EXERCISE 5.5
(Chapters 1 to 22)**

1 A sequence is formed by starting with 5 and adding 8 each time.

  a Write down the first eight terms.

  b What is the difference between the sum of the first four terms and
    the sum of the next four terms in this sequence?

2 Find

  a $4\frac{7}{10} + 3\frac{4}{5}$

  b $5\frac{1}{6} - 2\frac{2}{3}$

  c $6 \times 3 \times 4 - 7 \times 3$

  d $18 \div 3 + 3 \times 3 - 4 \div 2$

3 a Find, giving your answer in metres, 156 cm + 84 mm + 2.1 m

b Find, giving your answer in grams, 465 g + 0.44 kg + 750 mg

c Find, giving your answer in inches, 10 inches + 3 feet + 1 yard

4 a The product of two numbers is 2.478
If one number is 2.95, what is the other?

b A rectangular box measures 4.78 cm by 3.23 cm by 1.88 cm.
Estimate its capacity.
Use a calculator to find the capacity of the box correct to
2 decimal places.
How near is your estimate?

5 In a boat race one crew rows at 30 strokes to the minute while
another crew rows at 20 strokes to the minute.

a How long is it, in seconds, between strokes for
i the faster crew   **ii** the slower crew?

b They start the race by stroking at the same moment.
How long will it be before they stroke at the same instant again?

6 a By choosing a letter to stand for the number of length units of the
side of an equilateral triangle, find a formula for the perimeter,
P units of the triangle.

b The formula for changing x °C to y °F is $y = \dfrac{9x}{5} + 32$.

Find the temperature in °F when a thermometer reads 20 °C.

7 The bar chart shows how a family's weekly income is spent.

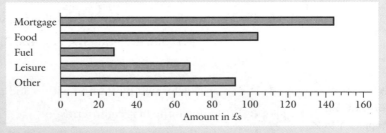

a On which item do they spend
i most   **ii** least?

b How much do they spend on
i fuel   **ii** leisure?

c How much do they spend altogether?

8 Solve the equations

   a $3x - 4 = 5$ **b** $2 - x = 4 + x$ **c** $0.7 = 0.9 - 4x$

9 The diagram shows a rectangular plot of ground measuring 30 m by 20 m. It is laid out as a rectangular lawn surrounded by a path that is everywhere 1 m wide.
Find

  a the measurements of the lawn

  b the area of the lawn

  c the area of the whole plot

  d the area of the path

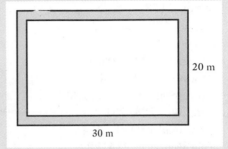

20 m

30 m

10 Hank has three pieces of string. The first piece is x cm long, the second piece is 4 cm longer than the first and the third piece is 5 cm longer than the second. Altogether he has 49 cm of string. How long is each piece?

11 The ages (in years) of 20 children attending a swimming class are

$$9, \quad 9, \ 10, \ 10, \ 10, \ 10, \ 10, \ 10, \ 10, \ 11,$$
$$11, \ 11, \ 11, \ 11, \ 11, \ 11, \ 11, \ 11, \ 12, \ 12$$

The ages (in years) of a different group of 20 children in a trampoline class are

  6, 7, 7, 8, 8, 8, 9, 9, 9, 9, 9, 10, 10, 11, 11, 11, 12, 13, 13, 14

  a Find the mean and range of the ages of each group.

  b Use your answers to part **a** to compare the ages of the two groups.

INDEX